THE
VEGENAISE®
Cookbook

FROM THE KITCHENS OF

Follow Your Heart

THE

VEGENAISE®

Cookbook

Great Food That's Vegan, Too

BOB GOLDBERG

The Countryman Press
A division of W. W. Norton & Company
Independent Publishers Since 1923

CONTENTS

HOW DO YOU SAY IT?

Is it "Veg'e-naise" or "Ve'gan-aise?"

I've been asked this question a thousand times, so I know it will come up for more than a few readers of this book. Over the years, I've heard the name of our product pronounced so many different ways, sometimes with hesitance, but just as often quite authoritatively. I've heard it all, including: "Vee'-guh-naise," "Vee-guh-neez," "Vej'-e-neez," and "Vee'-ja-neez." You name it, someone has said it.

So, is it VEGANaise, as the vast majority of people know it, or is it VEGE-naise?

We knew that the Federal Standard of Identity, a list of rules that the government has established to regulate the use of names of foods, prohibited us from using the word "mayonnaise" to describe a product which didn't meet a certain criterion which included eggs and sugar, both of which we do not use. Yet, we had to have a way to communicate to shoppers what the product was when they encountered it in a store. "Mayonnaise" was out, but the ending "naise" seemed to perfectly and uniquely tie it to that familiar product.

I started by making a list of all the names I could think of. I remember that Vegenaise was on the list, as well as Soyanaise, a name which fortunately was not chosen, because it would have become a problem years later when we decided to produce a soy-free version. Vegan-aise didn't make the cut because at the time the word "vegan" was not yet widely known. I first became familiar with veganism in the early 1970s, around the time that our product was created, but my sense was that it was still a fairly obscure term. To test my theory, I went to a bookstore to check a variety of dictionaries. I looked through editions from all the major dictionary publishers and to my surprise, the word "vegan" was absent from all but one, the *Random House Dictionary of the English Language Unabridged*. This, by the way, was a full 30 years after the word was first coined by Donald Watson, founder of the Vegan Society. If the word couldn't even be found in most dictionaries, I concluded, it was a pretty good bet that it would not be recognized by many people.

Follow Your Heart, our original market and café was, at the time, not entirely vegan, but strictly lacto-vegetarian, meaning we didn't use any meat, fish, fowl or eggs. We wanted a way to convey to people that the spread we were putting on their sandwiches and into other recipes met this standard, that it was a strictly **VEGE**table-based mayon**NAISE**. So, from those two words, the name was born.

To all those of you who have been calling our product "Ve'gan-aise" (because it is, after all, a vegan mayonnaise) we say, go ahead. That's a great name that we've even trademarked along with "Veg'e-naise" so people don't confuse us with other similar products. If we were creating it today, we would probably have called it that, but it's a little late in the game to be making such a change.

While the name may be confusing and has divided many who think they know the right way to say it, the genuine appreciation and affection for Vegenaise has proved it all worthwhile. So, our advice? Follow your heart. Call it whatever feels right to you. Just please be sure to enjoy It.

THE VEGENAISE STORY

Mayonnaise was first invented in 1756 by a French chef and, until recently, it had only been made with eggs. It was believed for a long time using eggs was the only way you could produce a mayonnaise-like spread. But the recipes in this book were all inspired by another wonderfully versatile sauce: Vegenaise. Created in the mid-1970s by Follow Your Heart, Vegenaise is entirely plant-based, making it a natural choice for anyone who cannot, or chooses not to, consume animal products. The development of Vegenaise is a story that—despite having lived it—still amazes me. I have been asked to tell this story countless times and I am thrilled to share my journey from the beginning.

Oh, What a Time It Was

This story begins in 1970 with the humblest of enterprises: a small, seven-seat vegetarian food bar, renting space in the back of an old-fashioned, '50s-era health-food store. Typical of the time, it catered to a relatively small number of people who were concerned with their health and who believed that consuming natural, less processed foods was a pathway to a healthier existence. Nothing about the place resembled the natural-food stores of today, with their huge selection and expansive size. The dimly lit store was partitioned from a single-story building; its 1,300 square feet taken up by two aisles of dry, often faded retail products, an aging produce case and freezer, a small stock room, and that seven-seat counter. There were two and a half employees: one who ran the store, another who operated the food bar, and a part-time person who came in three days a week to help put product on the shelves.

The only excitement in the place was coming from Michael Besançon and Caren Diem, two eager 20ish entrepreneurs, who had recently taken on the task of running the food bar following the departure of the previous operators. Together, they turned what had been an uninspired menu of tuna fish sandwiches and beef hamburgers on whole wheat buns into a lacto-vegetarian offering, as prescribed by the teachings of the spiritual path that the young couple had recently embarked upon. After a fateful hitch-hiking trip to Mount Shasta in Northern California, they began observing a diet excluding all meat, fish, fowl, and eggs. Completely unaware of any of these events and living just half a mile down the street, I could never have imagined how deeply I would soon be drawn into this lifestyle, or how it would one day become my life's work.

I first met Michael when I happened to walk into this same small health-food store and my hunger drew my attention to the food bar tucked in its back corner. At the time, the place, then named American Natural Foods, was essentially what we called a pill shop because the emphasis was on supplements rather than food. A glance at the menu that first day led me to an old favorite, a fresh banana milkshake, and this one was so good I knew I would be back for more. Over the ensuing weeks and months, as I moved on to the avocado and sprout sandwiches, the freshly made vegetable soups, and everything else they offered, it began to dawn on me that this new vegetarian diet had more to it than just taste. Michael and I spent countless hours discussing the many facets of living a life of harmlessness: the health aspects, the mental calmness, the environmental benefits. I became a regular patron of the store, and gradually began working there on occasion. One day, Michael came to me with a request. He and Caren wanted to travel to India to see their spiritual teacher, and he asked me if I might be willing to take care of the food bar for them while they were gone. It seemed like it would be a fun experience, so I agreed. They were gone for three months and by the time they returned I was in deep.

Over the next couple of years, while the health-food store remained relatively static, the food bar began to take off. Paul Lewin, an old college friend from Indiana, had moved to California and into my apartment. As we always seemed to need more help, Paul joined in to help out, along with Michael's friend, Spencer Windbiel. The four of us hit it off and complemented each other's skills in an extraordinary way. The place was becoming increasingly popular, continually adding more seats and more workers, many of whom came from a nearby communal house populated by more than a dozen of Michael's friends. And then one day, Michael brought word from the owner that the store was being put up for sale.

We discussed trying to find a way to buy the store, but we were four hippies with no money. Unwilling to give up however, we devised a plan. The amount we needed to raise was only $15,000 and the store was fully stocked, including a lot of vitamins that alone carried a retail value of about $100,000. The potential for this deal was almost too good to be true. Ultimately, Paul and I were each able to secure loans of $7,000 from our parents, Michael contributed his interest in the food bar, Spencer put in what cash he was able to come up with at the time, and we formed a partnership to buy the store.

Immediately, we eliminated all meat, poultry, and fish and changed the name of

the store to Follow Your Heart, to match the name we had already been using on the food bar. We adopted the name Follow Your Heart because it so well described what we were doing there: we were being true to our beliefs. We also thought that the name would resonate with people on a human level and encourage them to do the same. We knew that those words would be spoken time and time again in relation to the store, and we felt that just the energy of those words being repeated would lead to good things. The inspiration had come to us from a song of the same name performed by The Sons of Champlin and written by their guitarist, Terry Haggarty, to whom we will be forever grateful.

By 1974, when we acquired the store, the original seven-seat food bar had, by popular demand and with some creative rearrangement, become a 22-seat restaurant. One might think that having the kind of dietary restrictions that we put on ourselves would be a limitation, but just the opposite was true. Even back then, when avoidance of certain foods was not as commonplace as it is today, not serving meat actually attracted business. Some customers were vegetarians, but many more came just because the food was delicious.

A Secret Ingredient

One critical ingredient for us that simply couldn't be found in those days was mayonnaise without eggs. Fortunately for us, there was one. It was called Jack Patton's Lecinaise, so named because, according to the maker, the eggs had been removed and replaced with soy lecithin, a common emulsifying agent. This product, also quite popular in the store, was indeed our secret sauce. We used it on everything, and we used it by the ton, literally. At one point, we tallied up that we had purchased about 40,000 pounds of Lecinaise. It had all the creamy deliciousness of the most popular mayonnaises, but according to the label, it was egg-free, preservative-free, and sugar-free. Without it, it's fair to say that we wouldn't have done as well.

At one point, a rumor was circulating that Lecinaise was not actually eggless. Since many of our patrons relied on this product to meet their dietary preferences, we reached out to Patton for verification. He assured us that the product was completely authentic and labeled properly, and he provided us with a letter on his company letterhead verifying the accuracy of the label, re-asserting that the product was, in fact, eggless. But shortly thereafter, we received word that a special team from the California Department of Food and Agriculture had, during the night, raided a facility where Patton and his employees were soaking labels off jars of regular, store brand mayonnaise, and relabeling them with his own false claims. It turned out that our "secret ingredient" contained eggs, sugar, and preservatives. And all at once, we didn't have anything to use on our sandwiches.

Out of Crisis and Into Opportunity

We were in a bind to say the least. We tried doctoring up a tasteless, egg-free imitation mayonnaise by adding additional sweeteners and vinegar, but the resulting goop was terrible, and time was running out. We needed a delicious eggless mayonnaise to replace what we had lost, and I didn't know where to begin. Mayonnaise manufacturers that I questioned all insisted that there was no way to make mayonnaise without eggs. Clearly, this was the mindset and belief at the time. After decades of experience working in this field, I now know that there is much more to making mayo than meets the eye.

After many frustrating weeks trying fruitlessly to solve the eggless mayonnaise problem, one night during a dream, I was struck with what was, for me, an earth-shattering thought. The moment it happened was like the proverbial lightning bolt. From a sound sleep, I shot up to a sitting position in bed, attempting to focus on the vision which had come to me: the secret was not lecithin at all, but protein, a component far more abundant in eggs than lecithin.

It was about 3:00 in the morning when this revelation hit, and I was hardly able to contain myself until the morning light to get up and test my theory. I still wasn't quite sure which source to get the protein from, but I did have some tofu in the house which I thought I might try adding.

As the dawn approached, I set up my workspace on the kitchen counter and set to work. To my stunned amazement, success was immediate. It was far from perfect, but it was clear that I was on to something. There and then began the process of testing and refining that went on for months and numbered into the hundreds of trials. All my home refrigerator space was sacrificed to hold samples as it was not only necessary to keep each trial for comparison, but also to find out if the emulsions would remain stable over time.

During an earlier tour I had taken of a tofu factory, I saw that the manufacturing process produced many broken pieces of tofu. Those pieces were packed in bulk and sold to other companies that didn't need whole blocks. I reasoned that these pieces would be perfect for my purposes and they were cheaper as they were essentially waste. These pieces of tofu became the protein source that was used in the earliest versions of Vegenaise, the name that we eventually came to call this truly egg-free mayo.

I was able to make good Vegenaise in my blender at home but scaling up that process was a challenge. As the complexities of manufacturing became clearer, I realized that using fresh tofu as a protein source wasn't the most practical choice. I began to look for another solution. I knew that soy came in a wide variety of forms and there were hundreds of variations, each with their unique tastes and attributes. Some were gritty, some smooth, some darker, some lighter. What I learned was that each had properties, characteristics, and functions that were best for certain uses, but which would have to be tested in my application to find out how well they would work there.

Getting technical help from ingredient manufacturers was tricky. If they had known that I was developing this product so I could use it in my 22-seat vegetarian restaurant, that would have been the end for sure. These were multi-billion dollar food ingredient companies. Fortunately, some of these scientists were extremely generous with their time and knowledge, and without them, I don't know how I would have managed. The testing went on seemingly forever, until at last, a replacement for the tofu was found that functioned well as an emulsifier and imparted no negative characteristics.

Getting the flavor profile right became my next focus. I learned that the public's taste for mayonnaise had a regional component. Generally, Southerners preferred a sweeter product, more like Kraft's Miracle Whip. For me however, growing up in the Midwest, and for most consumers who live north of the Mason-Dixon Line, there was one brand that epitomized mayonnaise: Hellmann's Real Mayonnaise. Hellmann's has long been the best-selling mayonnaise in the United States, followed by Best Foods— said to be made using the same formula but produced and sold in the Western states. Still, for me, Hellmann's was the best, so that was the taste I was hoping to achieve. Getting one product to taste like another when pretty much all of the ingredients are different is an enormous challenge, but that was what I set out to do. To be sure my testing was true to the mark, I hand carried several jars of Hellmann's back to California with me each time I flew to Chicago for family visits. In the end, I think we came pretty close. I actually like Vegenaise better, and I've been told the same thing by fans of it for years. I never get tired of hearing that.

Having solved the first challenge of formulation, I now had to tackle how the product would be made. My plan was to find a co-packer, a manufacturer who makes products for other companies, who would agree to produce quantities small enough for our needs. This turned out to be difficult, however, as the economics favor very large orders and the situation was further complicated by the fact that we insisted on ingredients that most mayonnaise producers were unfamiliar with. Basically, everything about our product was different, from the need to keep it refrigerated to the lack of eggs, refined sugar, and the usual preservatives.

Ultimately, we found a company that was willing to pack Vegenaise into 5-gallon pails for us, but because we were still a small customer, production of our orders was often delayed, our calls weren't returned, and we would occasionally come dangerously close to running out of product. The friction this caused led us to continually look for better co-packing partners, and in the early years, we went through quite a few. As our restaurant business grew, we were able to make progressively larger orders, but as a single user, we were always going to be small potatoes.

Growth and Lessons Learned

The exceptionally good taste and texture of Vegenaise did not go unnoticed by our restaurant customers. There was still nothing close to it available to purchase in stores, so many customers begged and pleaded for us to sell it to them, which we did, by packing some for them in our take-out soup containers. As time passed, and some of those folks moved out of the area, they would sometimes write to us and ask if there was a way that we could send it to them. As there was really no practical method for

doing so, we began to think about the viability of packaging the product for sale and distribution across a broader geographic area. But how? Our little restaurant had no capital with which to launch a retail product and frankly we didn't have the vaguest idea how to go about it. All of this was new territory and back then there was no Google to go to for answers—for that matter, there wasn't even an Internet. Still, in time, we were able to figure out enough of the details and persuade our co-packer to produce a single pallet of jars for us. In mid-1977, our first full pallet of jars was produced and sold to a distributor. It was a time for celebration and the anticipation of what was to come had us over the moon. That was until the day that we got a call from the distributor that there was a problem.

Oil was leaking out of the jars. It was staining the paper labels, and there appeared to be some separation within the jars themselves. We had no idea what had gone wrong with that first retail run of Vegenaise, but we were crestfallen by the development, and what was worse for us at the time was that we had to buy it all back. Stung badly, we retreated to making it just for ourselves and didn't attempt to release it again for nearly 20 years.

One difference between this retail Vegenaise and what we had successfully been using in our restaurant without any problems was that, rather than being refrigerated, this packaged product was sold from the dry grocery shelf with all the other mayonnaise-type spreads. Ultimately, we concluded that without refrigeration, our product would not be suitable for distribution and that was the end of that.

Opportunity Knocks Again

In the fall of 1988, some of the salads that we made in our restaurant kitchen and sold in the store's deli case came to the attention of buyers at a local and rapidly growing grocery chain called Trader Joe's. They were looking for products to carry and thought ours would do well, and indeed they did. It wasn't long before it became clear that we would need to build out a separate facility just to handle their orders. We set up this new operation not far from the store and called it Earth Island, a name inspired by R. Buckminster Fuller's concept that we are all travelers on "Spaceship Earth."

With the cash flow provided by the growing business with Trader Joe's, we finally had the facility and the machinery to begin making and packaging our own Vegenaise. We knew what it would take to keep it stable, but we were concerned with how consumers would react to a mayonnaise-type spread that required refrigeration. Would they even know where to look for it in the store? It certainly couldn't be displayed next to the shelf-stable mayonnaise. We had learned that the hard way. Would the stores accept such a product? If not with the other mayonnaises, where would the retailers put it? In the refrigerator with cheese? There was no category for refrigerated mayonnaise at the time. We would need to create one. Despite these concerns, we decided to move forward and see what would happen.

Since our earliest iteration of Vegenaise, then made with almond oil, had been pulled from the shelves, a lot had happened in the vegetable oils market. Previously, most mayonnaise had been made with soybean oil, which was neutral in flavor, readily available, and relatively inexpensive. But in the ensuing years, a new oil from a plant commonly grown in Canada, called rapeseed, was bred to have lower amounts of erucic acid, making it suitable for human consumption. The benefit of rapeseed oil was that it had lower levels of saturated fats than any other vegetable oils, and this was considered important from a health standpoint. While still more expensive than soybean oil, rapeseed oil—now dubbed Canola oil, (an acronym for "CANada Oil Low Acid")—became a popular oil for dressings aimed at more health-conscious consumers. So, we decided to use Canola oil for the product we were planning to make for retail.

At about this time, I received a cold call from a gentleman who was selling something I had never heard of: grapeseed oil. He believed that his oil would make a great mayonnaise, and he was very persistent and convincing in telling me why. Grapeseed oil, not yet well known in the United States, had been shown in at least one study to have cardiovascular benefits, owing to a particular fatty acid profile which tended to lower total cholesterol and improve levels of HDL and LDL, seemingly leading to a reduction in cardiac events. This appeared to be a perfect fit for our product due to its association as a healthier alternative. So when we were finally ready to reintroduce retail Vegenaise, originally intended to be a single product, we ended up launching two varieties: Original and Grapeseed Oil. In the years that followed, many more variations have been added including Organic, Reduced-Fat, Soy-Free, Avocado Oil, Organic Garlic Aioli, and a number of other flavors, each designed to serve a different need or taste of our consumers.

A lot more has happened between that first production run of retail refrigerated product and today. We defined a new grocery category, created countless new products, line extensions, and new product innovations, but as this is *The Vegenaise Cookbook*, I will close with this: Today, Vegenaise is produced in our state-of-the-art, solar-powered Earth Island facility using the most sanitary, modern equipment and distributed in over 40 countries around the world. We prefer to make it in batches that are relatively small by industry standards, as opposed to the ultra-fast, continuous-process equipment used by the largest, mainstream brands. From the window next to my desk, which overlooks our Vegenaise production line, I'm still able to keep an eye on the process—although with today's quality checks in place, that's more for the satisfaction of watching the bottles whiz by than out of necessity.

Surrounded by an amazing, dedicated staff that has come together to make Vegenaise a successful and beloved product, and with constant reminders from consumers of how much they enjoy our products, I could not ask for more. In following my heart, I have been rewarded with a wonderful life and everything that goes along with that. I have been blessed in my family life, my business relationships, and with the good health with which to enjoy this life. To each and every person who reads this book, I wish the same. Enjoy the recipes and remember: Follow Your Heart.

The Party Is On

Roasted Red Pepper and Corn Bisque

SERVES 4 TO 6

In an honored tradition in our Follow Your Heart kitchen, we make our daily soups from scratch as we have since the early '70s. All four of the original owners were soup wizards and these soups became the day-to-day magnet to keep that early lunch crowd coming back. The freshly roasted red pepper and corn kernels blended with cumin and chili powder bring a smoky, rich kick, while the cilantro-lime Vegenaise crema helps cool it all down. The soup selection changes daily, so you have to keep coming back to try them all.

Soup

1 large red bell pepper, cut in half and seeded

1 tablespoon olive oil

3 cups corn kernels (from 4 ears of corn)

1 white onion, chopped

1 garlic clove, crushed and minced

2 cups veggie broth

2 teaspoons ground cumin

2 teaspoons chili powder

½ teaspoon cayenne pepper

Sea salt and freshly ground black pepper

Leaves from 2 sprigs cilantro

Cilantro-Lime Crema

1 cup Vegenaise

3 tablespoons fresh lime juice

¼ cup finely chopped fresh cilantro, plus more for garnish

Prepare the soup: Place the bell pepper, skin side up, on a baking sheet and coat with the olive oil. Broil in the oven for 10 to 15 minutes. Lower the heat to 250°F and cook for 5 minutes. Remove from the oven and set aside to cool. When cool enough to handle, scrape off the blackened skin.

Place the corn kernels in a large, dry skillet and roast over medium-high heat until they brown, stirring occasionally. Transfer to a bowl and set aside.

Place the onion in a large skillet and cook over medium heat for 3 to 5 minutes, until it begins to brown, stirring occasionally. Lower the heat and sauté over medium-low heat for 5 more minutes, or until translucent. Add the garlic and cook for 1 minute. Add the roasted corn and bell pepper and cook, stirring occasionally, for an additional 5 minutes. Remove from the heat and, reserving some of the mixture for garnish, transfer to a blender.

Add the veggie broth, cumin, chili powder, and cayenne to the blender. Remove the middle section of the lid and place a clean dish towel over the hole to prevent splatter. Blend on high speed until smooth. Add salt and black pepper to taste.

Prepare the Cilantro-Lime Crema: Combine the Vegenaise, lime juice, and cilantro in a bowl, using a fork or whisk.

Serve the bisque with a drizzle of Cilantro-Lime Crema, and garnish with cilantro and the reserved roasted corn and pepper.

Crunchy Fried Okra with Rémoulade

Many people aren't very familiar with okra and we used it in limited restaurant recipes back in the '80s, usually in a gumbo or stew. By cutting okra on the bias, you really showcase the beauty of the pods, and by dredging the okra in a Vegenaise and vinegar blend, you help coat them and prepare them for frying. Plus, the smoky rémoulade complements the hearty crunch and savory flavor of the fried okra and ensures a happy, snacking crowd.

SERVES 2 TO 4

8 ounces fresh okra (about 15 pods)

½ cup Vegenaise

¼ cup cider vinegar

1 cup cornmeal

¼ cup unbleached all-purpose flour

½ teaspoon sea salt

¼ teaspoon freshly ground black pepper

2 cups neutral-tasting oil (safflower or grapeseed) for frying

½ cup Rémoulade (page 155) for serving

Trim the stems of the okra and cut on the bias into bite-size pieces.

Place the Vegenaise in a medium bowl and whisk in the vinegar until thinned and smooth. Immerse the okra in the Vegenaise mixture and let marinate for 20 minutes.

Stir together the cornmeal, flour, salt, and black pepper in a medium bowl. Place the okra bowl next to the cornmeal bowl. Before transferring the okra to the cornmeal bowl, allow most of the liquid to drip back into the okra bowl, then dredge with the cornmeal mixture until evenly coated. Place the coated okra on a parchment-lined baking sheet.

Heat the oil in a Dutch oven or cast-iron skillet over medium-high heat. When the oil is hot (350°F), use a slotted metal spoon to drop one piece of okra into the oil. The okra should sizzle immediately, and if it doesn't, increase the heat. Cook the okra in batches for 3 minutes per batch, or until just golden brown. Transfer to a paper towel–lined wire rack to cool slightly and absorb any excess oil.

Serve with the Rémoulade.

Bob Says

In California, fresh okra season is July through September, so this is a perfect summer snack. If you can't get fresh okra, frozen works; just be sure to thaw ahead of time on a paper towel so the liquid is absorbed.

Lemon & Thyme Broccolini Tarts

These bright and savory tarts were inspired by my daughter Claire's asparagus tart recipe that we look forward to in early spring. We gather fresh lemons from our tree and the light addition of thyme balances the subtle char of the broccolini. One variation she makes is to use only the florets and cut the pastry into 2-by-2-inch bites, making an excellent appetizer. And don't forget to save your stems for a freezer broth base.

SERVES 2 TO 4

1 pound broccolini

¼ cup Vegenaise

¼ cup fresh lemon juice

1 tablespoon chopped fresh thyme

1 garlic clove, crushed and minced

2 teaspoons freshly ground black pepper

1 sheet puff pastry, fully thawed and chilled in the fridge

All-purpose flour for dusting

2 tablespoons vegan butter, melted

1 lemon, sliced into thin rounds

¼ teaspoon red pepper flakes

2 tablespoons Follow Your Heart Shredded Parmesan

¼ cup toasted pine nuts

2 teaspoons lemon zest

Preheat the oven to 375°F. Line a baking sheet with parchment paper.

Trim off the ends of the broccolini. Slice any thick stalks in half lengthwise so all the stalks are about the same thickness.

Whisk together the Vegenaise, lemon juice, thyme, garlic, and black pepper in a large shallow baking dish. Add the broccolini and coat. Marinate for 10 minutes.

Roll out the thawed puff pastry on a lightly floured surface into an 8-by-14-inch rectangle. Cut into quarters and transfer to the prepared baking sheet. Lightly score a ½-inch border around the edges of each quarter with a knife and poke the centers with a fork. Lightly brush edges with the melted vegan butter.

Slice the broccolini into pieces that are equal in length to the inner border of your pastry's scored lines. Position the broccolini across the inside of the puff pastry, alternating stems and florets. Layer with lemon slices to cover the broccolini. Sprinkle with the red pepper flakes and Parmesan. Bake for 20 to 25 minutes, until the crust is golden brown and the broccolini is slightly crispy.

Garnish with the pine nuts and a sprinkle of lemon zest.

Crispy Purple Rice Cakes with Edamame Vegenaise

These gorgeous rice cakes were originally brought to us through Danny Seo and a chef he introduced us to named Olivia Roszkowski. Danny really inspired us to complete this book and I'm very grateful for his expertise, insights, and most importantly, his friendly push. Olivia has a talent for bringing an elegance to simple ingredients and the subtle flavor of the wasabi in the rich edamame Vegenaise brings their flavors to the front of each bite.

YIELDS 12 RICE CAKES AND 1 CUP EDAMAME VEGENAISE

Rice Cakes

1 cup uncooked purple rice

½ teaspoon sea salt

1¾ cups water

Edamame Vegenaise

2 cups fresh mint (1 small bunch)

1 cup shelled edamame, plus more for garnish

½ cup Vegenaise

2 teaspoons wasabi powder

¼ cup filtered water

½ cup sesame seeds

Preheat the oven to 350°F. Line a baking sheet with parchment paper.

Prepare the rice cakes: Combine the rice, ¼ teaspoon of the salt, and the 1¾ cups of water in a medium saucepan and bring to a simmer. Bring the heat to low, cover, and cook for 30 minutes. Remove from the heat and set aside to cool.

Remove the leaves from the mint sprigs, reserving a few whole sprigs for garnish. Place the leaves in a fine-mesh strainer and blanch for 5 seconds by holding the strainer over the sink and pouring hot water over the leaves.

Prepare the Edamame Vegenaise: Combine the blanched mint, edamame, Vegenaise, wasabi powder, remaining ¼ teaspoon of salt, and the filtered water in a blender or food processor and blend until smooth.

Toast the sesame seeds in a skillet over high heat for 30 seconds, then stir into the cooled rice.

Form cakes from the rice mixture, using a ¼-cup measure, and flatten them to ½ to ¾ inches in height and 3 inches in diameter. Place on the prepared baking sheet and bake for 20 minutes, or until crisp.

Top with the Edamame Vegenaise, and garnish with extra edamame and the reserved mint sprigs.

Molten Hearts of Palm "Crab" Dip with Eggplant Chips

I hope I'm not taking a risk with my storybook marriage by revealing this, but two of my wife Kathy's favorite foods are eggplant and potato chips. (Admittedly, the potato chips are only an occasional guilty pleasure and must be of the highest quality. No ordinary chips will do.) These eggplant chips, however, are the perfect marriage of her favorites, and the spicy and savory harissa "crab" dip couldn't be a better complement to these simply sautéed eggplant chips.

YIELDS 1½ CUPS DIP
AND 25 TO 30 CHIPS

"Crab" Dip

One 14-ounce jar hearts of palm

¼ cup Vegenaise

1 teaspoon harissa paste (page 171 or store-bought), or to taste, plus more for garnish

1 scallion, green part only

¼ teaspoon sea salt

Eggplant Chips

1 large eggplant

½ teaspoon sea salt

¼ cup olive oil for frying

Preheat the oven to 450°F.

Prepare the "crab" dip: Drain and rinse the hearts of palm and blot dry. Shred with a fork and place in a medium bowl. Add the Vegenaise and harissa and toss to combine.

Slice and chop the scallion (approximately 2 tablespoons) and, reserving a few pieces for garnish, stir into the hearts of palm mixture. Season with ¼ teaspoon of the salt.

Transfer the hearts of palm mixture to a 9-by-13-inch baking dish and bake for 20 minutes, or until molten and slightly golden. Top with the reserved scallion pieces and a dollop of additional harissa.

Prepare the eggplant chips: Slice the eggplant into ¼-inch-thick rounds. Place in a colander and sprinkle with the ½ teaspoon of salt. Allow the eggplant to sit for 10 minutes, then squeeze out any excess moisture from each slice.

Heat the oil in a nonstick or cast-iron skillet over medium heat and cook the eggplant slices for 1 to 2 minutes per side, until golden.

Serve the eggplant chips with the warm "crab" dip.

Sonora "Ceviche"

The flavors in this ceviche remind me of some fun and adventurous times that Paul and I had while in Mexico, back in the '70s and '80s. I'm not quite ready to share all the details of those adventures, but they did inspire this recipe. Although we've been using these "tree vegetables" for years, hearts of palm are an up-and-coming fish alternative. They're the meaty core of some varieties of palm trees, and are most commonly harvested in such places as Costa Rica and Hawaii. This recipe uses them to mimic the method of "cooking" seafood (typically shrimp or scallops) in a marinade of lime or lemon juice, known as ceviche. And don't shy away from the lime—it's a core ingredient to replicate that traditional ceviche flavor.

SERVES 4 TO 6

Ceviche

One 14-ounce jar hearts of palm

½ cup fresh lime juice

¼ cup Vegenaise

½ teaspoon sea salt

1 teaspoon freshly ground black pepper

1 small jalapeño pepper, seeded and finely diced

½ cup diced fresh mango

½ cup diced peeled cucumber

½ cup seeded and diced cherry tomato

¼ cup sliced unpeeled radish

¼ cup diced red onion

3 tablespoons chopped fresh cilantro leaves

2 firm, ripe avocados

One 13-ounce bag fresh tortilla chips

Vegan "Clam" Juice

¼ cup tomato juice

1 tablespoon liquid aminos

2 teaspoons vegan Worcestershire sauce

Drain and rinse the hearts of palm and blot dry.

Combine ¼ cup of the lime juice with the Vegenaise, salt, and black pepper in a large bowl. Immerse the hearts of palm in the mixture and marinate for at least 15 minutes in the refrigerator.

Meanwhile, prepare vegan "clam" juice: Whisk together the tomato juice, liquid aminos, and Worcestershire sauce in a small bowl and set aside.

Pull the hearts of palm from the fridge and strain out the liquid. Add the mango, cucumber, tomato, radish, red onion, jalapeño, and cilantro. Gently toss to coat and refrigerate for another 30 to 45 minutes. Strain once more and add the vegan "clam" juice.

Just prior to serving, chop the avocado and coat with the remaining ¼ cup of lime juice to prevent browning. Gently fold the avocado into the ceviche.

Serve with the tortilla chips.

Crispy Artichokes with Lemon-Pepper Aïoli

Toward the end of every summer, Paul and I, along with our families, pack up and head to Hawaii for vacation. This is a chance for us to unwind and enjoy our time together. On our must-visit list is a favorite shack of a bar that has the best crispy artichokes, and this recipe is inspired by those crunchy nuggets. This lemon pepper aïoli is excellent with steamed artichokes, but the indulgence comes from frying them and yields an irresistible outcome.

SERVES 2 TO 4

Lemon-Pepper Aïoli

1 cup Vegenaise

2 garlic cloves, crushed and minced

1 teaspoon lemon zest

2 tablespoons fresh lemon juice

1¼ tablespoons freshly ground black pepper

Artichokes

½ cup Vegenaise

¼ cup unsweetened almond milk

½ (4-ounce) package Follow Your Heart Shredded Parmesan

½ teaspoon garlic powder

1½ cups vegan panko bread crumbs

1 tablespoon minced fresh parsley, patted dry

½ cup unbleached all-purpose flour

One 15-ounce can quartered artichoke hearts, drained and patted dry

3 cups neutral-tasting oil (safflower or grapeseed) for frying

¼ cup Follow Your Heart Grated Parmesan for sprinkling

Prepare the Lemon-Pepper Aïoli: Combine the Vegenaise, garlic, lemon zest, lemon juice, and black pepper in a small bowl. Set aside in the refrigerator for at least 15 minutes.

Prepare the artichokes: Place the Vegenaise in a medium bowl and whisk in the almond milk until smooth.

Mix together the shredded Parmesan, garlic powder, panko bread crumbs, parsley, and flour in a small bowl. Position this bowl next to the Vegenaise bowl.

Dredge the artichokes in the Vegenaise mixture, then switch hands and coat the artichokes in the bread crumb mixture. Place the coated artichokes on a baking sheet.

Heat the oil in a Dutch oven or cast-iron skillet over medium-high heat. When the oil is hot (350°F), use a slotted metal spoon to fry the artichokes in batches. Cook until golden brown and crunchy, 3 to 4 minutes. Transfer to a paper towel–lined wire rack to cool slightly and absorb any excess oil.

Sprinkle with the grated Parmesan and serve with the Lemon-Pepper Aïoli.

Roasted Elotes (Mexican Street Corn)

If corn is on the menu, I'm guaranteed to order it. Here in So Cal, the summers are hot, but the evenings cool down to a temperature where your skin and the air are the same. For these evenings, we often find ourselves firing up our backyard grill with fresh, local corn on the cob. These elotes are summer in a cob.

SERVES 6

Avocado-Cilantro Crema

1 ripe avocado

½ cup Follow Your Heart Dairy-Free Sour Cream

¼ cup packed fresh cilantro

Juice of ½ lime

Corn

6 large ears of fresh corn, husks and silk removed

2 tablespoons vegan butter, melted

1 tablespoon sea salt

½ cup Vegenaise

1 cup Follow Your Heart Grated Parmesan, plus more for garnish

¼ cup chopped fresh cilantro, plus more for garnish

½ cup Chipotle Vegenaise

2 limes, cut into wedges, for serving

Prepare the Avocado-Cilantro Crema: Combine all the crema ingredients in a food processor and process until smooth and creamy.

Prepare the corn: Preheat a grill or grill pan to medium-high heat.

Brush each cob lightly with the vegan butter and add a light sprinkle of salt. Grill until evenly charred, turning often, about 10 minutes. Alternatively, you can roast the corn in a 400°F oven for 20 to 25 minutes; or bring a large pot of water to a boil, add the corn, bring the water to a boil again, turn off the heat, and let the corn cook for 3 to 5 minutes.

Lightly brush each cooked ear of corn with Vegenaise, then sprinkle with the Parmesan and cilantro. Serve the corn with a drizzle of Chipotle Vegenaise and the Avocado-Cilantro Crema. Garnish with the lime wedges and an extra sprinkle of Parmesan and cilantro.

Herbed Parmesan Polenta Fries

In our house, we love polenta. While many people often enjoy it as a creamy, grits-like side, we prefer it crispy. Kathy's polenta bakes are a labor of love that steal the show every time. Serve these polenta fries with our Basic Garlic Aïoli (page 147), or even a simple pesto sauce or classic marinara.

SERVES 4 TO 6

1 cup uncooked polenta or corn grits

1 teaspoon chopped fresh sage

1 teaspoon chopped fresh basil

1 teaspoon sea salt

3½ cups water

2 tablespoons vegan butter

½ cup Follow Your Heart Shredded Parmesan

1 tablespoon olive oil for pan and brushing

Basic Garlic Aïoli (page 147)

Combine the polenta, herbs, salt, and water in a medium saucepan and bring to a boil while whisking. Lower the heat to a simmer, then cook for 10 to 15 minutes while stirring constantly so there are no lumps.

Once the polenta mixture begins to thicken and pull away from sides of the saucepan, stir in the vegan butter and Parmesan. Transfer to an 8-inch square baking dish and spread out in an even layer. Let cool, uncovered, in the refrigerator for 45 minutes to 1 hour, until set.

Preheat the broiler.

Cut the polenta into 4-by-1-inch sticks. Line a baking sheet with foil and brush with the oil. Arrange the polenta sticks in a single layer on the prepared baking sheet and brush their tops with the oil. Broil for 15 to 20 minutes, until golden brown, flipping once halfway through. Transfer to a paper towel–lined wire rack to cool slightly and absorb any excess oil.

Serve with the Basic Garlic Aïoli.

Sweet Potato Oven Fries with Curry Mayo

A few years ago, we added sweet potato fries to the Follow Your Heart menu as an option alongside our regular potato fries. Something about the richness of the sweet potatoes really seems to play well with the creamy texture of Vegenaise. Once you try a sweet potato fry with our Curry Mayo, which is by far the hero of this dish and something that I rave about every time I taste it, you'll never go back!

SERVES 2 TO 4

2 large sweet potatoes, peeled and julienned

¼ cup olive oil

Sea salt and freshly ground black pepper

½ cup Curry Mayo (page 150)

Preheat the oven to 425°F.

Dunk the sweet potatoes in an ice bath for roughly 1 minute and pat dry.

Combine the sweet potatoes, olive oil, salt, and black pepper in a nonreactive bowl (e.g., glass or stainless steel). Arrange in a single layer on a baking sheet. Bake for 15 minutes, then flip and bake for an additional 10 to 15 minutes, until sizzling and edges begin to brown.

Remove the fries from the oven and serve with the Curry Mayo.

Grilled Vegetable Corn Cakes with Rémoulade

I simply can't resist ordering corn cakes when I see them on a menu. So, it's not surprising that we often served fried corn fritters as a special in the Follow Your Heart Restaurant. This, in turn, inspired the Vegenaise-based Tartar Sauce that we launched back in 2009. At home, though, you can opt for this pan-friendly version of the cakes. Fresh corn kernels are by far the best choice, but in a pinch, you can use frozen or canned. Just be sure to defrost and fully drain them.

YIELDS 6 TO 8 CAKES

2 teaspoons olive oil (total) to sauté corn and fry cakes

½ cup corn kernels (fresh, frozen, or canned)

¼ cup seeded and diced mixed red and green bell pepper

1 teaspoon sea salt, plus more for seasoning

Freshly ground black pepper

½ to 1 cup yellow cornmeal

½ cup corn grits

¼ cup evaporated cane sugar

5 cups water

¼ cup vegan butter

1 tablespoon chopped fresh cilantro

½ cup Rémoulade (page 155)

Lightly oil a skillet and place over medium-high heat. Sauté the corn and diced peppers until the corn kernels are lightly browned. Season with salt and black pepper to taste and set aside.

Stir together the ½ cup of the cornmeal, corn grits, and sugar in a medium bowl.

Bring the water to a boil in a medium saucepan. Add 1 teaspoon of the sea salt and lower the heat to a simmer. Slowly (to avoid lumps), stir the corn-meal mixture into the simmering water, stirring constantly, until the corn mush becomes thick enough that a wooden spoon will remain upright when placed in the middle. Add more cornmeal, ¼ cup at a time, if too thin. Add the vegan butter to the corn and pepper mixture in small pieces and stir until melted. Add the corn and pepper mixture and cilantro.

Spoon into 2½-by-¾-inch stainless-steel ring molds or form by hand into cakes and set on a sheet of waxed paper. Refrigerate, uncovered, for a mini-mum of 1 hour, to firm up.

Remove from the molds and cook in a lightly oiled large skillet over medium heat until golden brown, 8 to 10 minutes, flipping halfway through.

Serve with the Rémoulade.

Spaghetti Squash Taco Shells with Spring Pico

I always cringe when I need to cut through a large squash or gourd, for fear of losing a finger while I wobble my way through the thick skin and girth of the squash. A trick I've learned from my daughter Claire is to microwave the squash for a few minutes to soften it, or you can shave off and flatten a side of the skin so it rests flat on your cutting surface. The effort, though, is worth it, as these taco shells are a deliciously vegetable-rich alternative that can be served as a classic taco or tostada base.

YIELDS 12 SHELLS AND 2 CUPS PICO

Tacos

1 small spaghetti squash

2 cups finely ground cornmeal

½ cup Vegenaise

Spring Pico

4 large tomatillos, husks removed

1 small bunch cilantro (about 1 cup)

2 garlic cloves

1½ teaspoons sea salt

Preheat the oven to 375°F. Line a baking sheet with parchment paper.

Prepare the tacos: Cut the spaghetti squash in half lengthwise, scoop out and discard the seeds, and place, cut side down, on the prepared baking sheet. Place the tomatillos on the same pan. Roast for 30 minutes. Carefully remove the tomatillos and set them aside to cool, then roast the squash for an additional 15 minutes, or until tender. Remove from the oven and set the squash aside to cool.

Prepare the Spring Pico: Stem the tomatillos and place in a blender. Remove the leaves from the cilantro, reserving a few sprigs for garnish. Add the leaves to the blender along with the garlic and 1 teaspoon of the sea salt. Blend until smooth.

To prepare the taco shells, use a fork to scrape cooked squash flesh into strands (about 3 cups). Combine the scraped squash, cornmeal, Vegenaise, and remaining ½ teaspoon of salt in a medium bowl.

Use a ¼-cup measure to form the mixture into tortillas, using your hands to flatten them into ⅛-inch-thick disks. Place on a freshly parchment paper–lined baking sheet.

Increase the oven temperature to 425°F and bake for 25 to 30 minutes, or until dry and golden. If you desire, toast in a cast-iron skillet over high heat after baking, for a more charred tortilla.

Top the baked taco shells with the reserved cilantro sprigs and serve with the Spring Pico.

BETWEEN TWO BREADS

The Avocado, Tomato, and Sprouts

The sandwich that started it all. Without the popularity of this sandwich and the abrupt loss of the allegedly egg-free mayo we were buying at the time—which later was found to be a fraud—we would never have Vegenaise today. There's a whole lot more to the story, which you'll find in "The Vegenaise Story" on page 8 of this book. But if I go back to what made this sandwich so popular, fresh ingredients were key: a bright red tomato with perfectly ripe avocado slices, and hearty wholegrain bread. This sandwich was so popular that in our tiny restaurant, we literally went through half a ton of avocados every week. They had to be hand sorted to confirm they were ripe and this was a daily task that my friend Spencer took on. He was *the* avocado pro. Without careful sorting, we would never have been able to deliver perfect avocados every time. Avocados don't begin to ripen until picked, but once that has occurred, the time until they are ready to use can vary widely. Early in the season, that could be as much as two to three weeks, whereas late season picks can be ready in just a few days. It doesn't get much better than The Avocado, Tomato, and Sprouts, which you can still order today.

YIELDS 2 SANDWICHES

4 slices vegan wholegrain bread

¼ cup Vegenaise

2 ripe avocados, pitted, peeled, and thinly sliced

½ cup sprouts, rinsed and patted dry

1 ripe tomato, sliced

Spread a good amount of Vegenaise on one side of all four slices of bread. On one slice, Vegenaise side up, arrange the avocado slices in an even layer, covering the entire piece of bread. Add a small handful of sprouts, then place two slices of tomato. Close the sandwiches and serve.

Bob Says

In addition to perfectly ripe avocados and bright red tomatoes, a key ingredient to this sandwich is the bread. It's worth the effort to get a freshly baked farmers' market loaf that's both soft and hearty.

Crispy Mushroom Sourdough Patty Melt

YIELDS 2 SANDWICHES

In the early days of the restaurant, we had a sandwich that we called the Barangrill. If you were born before 1980, you might recognize this as a reference to the Joni Mitchell song, often heard playing in the background at the restaurant and which inspired our version of the roadside diner classic. This modern riff on that recipe is very simple but so amazingly savory. By slow-cooking the mushrooms, you bring out their unique umami and earthy flavor. I'd love to show you the way to this tasty café standard.

2 large portobello mushrooms

2 tablespoons olive oil

1 teaspoon garlic powder

½ teaspoon sea salt

⅓ cup Vegenaise

4 slices vegan sourdough bread

1 cup baby arugula

Clean the mushrooms with a damp paper towel, remove the stem, and scrape out the gills. Slice the caps into thin strips.

Heat the olive oil in a large skillet over medium heat. Add the mushrooms and cook for 5 minutes, stirring occasionally. Stir in the garlic powder and salt and cook for an additional 10 minutes, or until golden. Remove the mushrooms from the skillet (you should have approximately 1 cup) and wipe out the skillet.

Spread the Vegenaise evenly on both sides of each slice of bread.

Place ½ cup each of the arugula and the mushrooms between two slices of bread to form each sandwich.

Reheat the skillet over medium heat. When hot, place both sandwiches onto the hot surface and top with a heavy weight or pan. Sear for 3 to 4 minutes on each side, or until the bread is golden and crispy.

Remove from the skillet, cut the sandwiches in half, and serve warm.

Portobello Banh Mi with Pickled Veggies

Did you know that *banh mi* actually means "sandwich"? So, a banh mi sandwich is a sandwich sandwich. I read that as an ultimate sandwich and this Vietnamese standard certainly delivers a powerful punch of flavor and may well deserve that extra emphasis. We recently added a version of this banh mi to our restaurant menu and it's become one of my go-to items. And I can never resist adding a little extra dollop of our bold Sriracha Vegenaise.

SERVES 4

Pickled Veggies

½ cup white vinegar

¼ cup evaporated cane sugar

½ teaspoon sea salt

2 medium carrots, julienned

⅓ daikon radish, julienned

½ English or Persian cucumber, julienned

Banh Mi

4 large or 6 medium portobello mushrooms

2 tablespoons dark agave nectar

1½ tablespoons tamari

1 tablespoon sriracha

3 garlic cloves, crushed and minced

2 tablespoons warm water

Sea salt and freshly ground black pepper

1 tablespoon olive oil

4 vegan Vietnamese-style baguettes or crispy French rolls

½ cup Sriracha Vegenaise

¼ cup fresh cilantro leaves

¼ cup fresh mint or Thai basil

Thin slices of jalapeño or chile pepper

Prepare the Pickled Veggies: Whisk together the vinegar, sugar, and salt in a medium, shallow bowl until the sugar is dissolved. Place the julienned veggies in the bowl and toss to coat and submerge them. Cover and refrigerate for at least 30 minutes. Drain before use.

Prepare the Banh Mi: Clean the mushrooms with a damp paper towel, remove the stem, and scrape out the gills. Cut the mushroom caps into quarters, poke all over with a fork, and set aside.

Whisk together the agave nectar, tamari, sriracha, minced garlic, and warm water in a large, shallow bowl. Add salt and black pepper to taste. Add the portobello quarters to the sriracha mixture and let marinate for 30 minutes, flipping over occasionally for an even soak.

Heat the oil in a large sauté pan over medium heat. Place the portobello quarters in the pan and allow them to brown on one side, then flip over to allow even browning on all sides. Pour half of the sriracha marinade into the pan and simmer until reduced by at least half, while turning the portobello quarters to prevent from burning. Once the mushrooms are tender, remove from the pan and set aside.

Build your banh mi by slicing the baguettes in half lengthwise. Spread the Sriracha Vegenaise on both sides of all the bread. Divide the pickled slaw among the bottoms of all four pieces of bread, top with the portobellos, then add the cilantro, mint, and sliced pepper. Close the sandwiches and serve.

Avocado Toast with Tomato Confit and Pickled Kale

At home, we've been eating some version of avocado toast for breakfast since the '70s, long before it was a "thing." This one is a home kitchen showstopper. Blistered cherry tomatoes, vibrant carrot and kale, and a touch of herbs all combine into a gorgeously colorful and delicious treat.

SERVES 2

1 cup whole cherry tomatoes

2 garlic cloves, crushed and minced

1 cup olive oil

1 teaspoon kosher salt

1 teaspoon dried thyme

1 teaspoon dried basil

¼ cup cider vinegar

1 tablespoon evaporated cane sugar

1 cup stemmed and chopped kale

½ cup shaved carrot (use a vegetable peeler)

2 slices your choice of vegan bread

2 tablespoons Vegenaise

½ large avocado, pitted, peeled, and sliced

Juice of ½ lemon

Sea salt and freshly ground black pepper

Preheat the oven to 300°F.

Toss the cherry tomatoes and garlic in the olive oil. Spread out in a single layer in a roasting pan. Sprinkle with the kosher salt, thyme, and basil.

Roast in the oven for about 1 hour, then remove from the oven and allow to cool.

Whisk together the cider vinegar and sugar in a medium bowl until the sugar is dissolved.

Toss the kale and carrot in the vinegar mixture and mix to coat. Gently massage and let sit in the marinade for at least 30 minutes.

Toast the bread and coat one side of each slice with the Vegenaise. Top the coated side with the avocado slices and mash with fork, leaving some chunks.

Sprinkle the lemon juice on the avocado; top with the kale slaw and then the cherry tomatoes. Sprinkle with salt and black pepper to taste.

Shredded Carrot and Potato Burgers

When we first added the Nutburger to our menu in the '70s, it was an instant hit. We shortly thereafter made it "Supreme" by adding fresh shredded carrot, lettuce, mushrooms, and sauerkraut. The result was indulgent and provided an added crispy bite that inspired this supreme take on a more traditional burger. The carrot and potato strips bring an unexpected crunch and then our American Slices add a creamy warmth that just melts your soul.

SERVES 4

4 large carrots, peeled

2 medium russet potatoes, peeled

2 tablespoons high-heat oil (safflower or canola)

Sea salt and freshly ground black pepper

4 vegan burgers of your choice

4 slices Follow Your Heart American Slices

4 vegan hamburger buns

¼ cup Vegenaise

1 ripe tomato, sliced

Using a mandoline or the shredding attachment of a food processor, separately shred the carrots and potatoes to about matchstick size. Quickly transfer the potato pieces to a large bowl filled with ice water. Let sit for a minute, then drain and pat dry.

Heat 1 tablespoon of the oil in a large skillet over medium-high heat. Add the potatoes, sprinkle with a pinch each of salt and black pepper, and cook about 4 minutes, then flip and cook until crispy and fried. Transfer to a paper towel–lined wire rack to drain.

Heat the remaining tablespoon of oil in the same skillet over medium-high heat. Add the carrots, sprinkle with a pinch each of salt and black pepper, cook for about 3 minutes, flip, and cook until nice and crispy. Transfer to same rack as the potato shreds and allow to drain.

Cook the burger patties according to their instructions. After flipping, add the American Slices and cover. Cook until heated through.

Toast the hamburger buns lightly, then slather with equal amounts Vegenaise on both cut sides. Assemble each burger in this order: bottom bun, patty, potato and carrot mixture, tomato, and top bun.

Creamy Salsa Verde Chick'n Grilled Cheese

You know you're lucky when you find employees who are talented and aligned with the mission of the company. You're even luckier when they turn out to be *multi*talented and even *more* lucky when they seemingly do it all. Oscar heads up our social media, does much of our photography, and happens to be pretty handy in the kitchen. He introduced me to simmering the chick'n in broth with the salsa verde so it infuses a tangy, rich flavor while maintaining a succulent texture that pairs nicely with the spice of our Pepper Jack Slices. For extra kick, you can add a few slices of jalapeño to the sauté broth; just check first to make sure everyone will appreciate it "picante."

SERVES 2

12 ounces vegan chicken strips

2 tablespoons olive oil

¼ cup veggie broth

1 cup salsa verde, plus more for serving

1 cup unsweetened nondairy creamer

Salt and freshly ground black pepper

7 Follow Your Heart Pepper Jack Slices

4 slices vegan sourdough bread

¼ cup Vegenaise

Tortilla chips for serving

Salsa verde for serving

Cut the vegan chicken strips into 1-inch cubes. Heat the olive oil in a large skillet over medium heat. Add the vegan chicken cubes and cook until slightly browned on all sides, about 5 minutes. Remove from the skillet and set aside.

Add the broth, salsa verde, creamer, salt, and black pepper to the skillet. Stir to combine, bring to a boil, cover, lower the heat, and simmer for about 15 minutes, until bubbling and well seasoned. While simmering, shred five slices of the Pepper Jack and set aside.

Add the vegan chicken and shredded Pepper Jack to the skillet and stir. Cook for additional 5 minutes, or until the Pepper Jack is melted.

Lightly coat both sides of the bread with Vegenaise. On one side of each sandwich, layer one slice of the remaining Pepper Jack and top with the vegan chicken mixture. Close the sandwiches; each will be lightly coated with Vegenaise on the outside.

Heat a large skillet over medium-high heat. Place the sandwiches in the skillet and cook about 3 minutes each side, or until bread is golden brown. Cut in half and serve with chips and salsa verde.

Lip-Smackin' BBQ Sandwich

Over the years, a few barbecue sandwiches have made it to our menu, but they come and go with the seasons. This one is an option all year round. A secret benefit to Vegenaise is how you can infuse the flavors of your favorite sauce into your vegetable or protein without having it stick to the pan or dry out before it's fully cooked. When we first launched the gourmet line of flavored Vegenaise, we played with different ways to use it, and this method became an instant hit. By simmering vegetables and seitan in your favorite barbecue sauce mixed with a little Vegenaise, you ensure rich taste and a juicy texture.

SERVES 4

½ cup Vegenaise

¼ cup barbecue sauce

½ yellow onion, sliced into ¼-inch strips

1 sweet red bell pepper, cut into ¼-inch strips

8 ounces seitan or other veggie meat, cut into ½- to- 1-inch strips

2 garlic cloves, crushed and minced

Splash of soy sauce for deglazing (optional)

Sea salt and freshly ground black pepper

4 vegan sandwich rolls

Deli mustard

Pickle slices

1 small red onion, thinly sliced

Heat a heavy-bottomed nonstick skillet over medium-high heat. Add ¼ cup of the Vegenaise and the barbecue sauce and sauté the yellow onion and bell pepper, stirring often, until the onion is slightly translucent. Add the veggie meat and cook until just browned. Lower the heat and simmer for 5 to 7 minutes, until bubbling and well seasoned. Add the garlic and cook for 1 minute more.

Turn off the heat and transfer the mixture to a bowl. If deglazing the pan, add a splash of soy sauce.

Season with salt and black pepper.

Serve on sandwich rolls spread with the remaining Vegenaise and deli mustard, and topped with pickles and red onion.

Marinated Kale and Avocado Sandwich

For years, the Follow Your Heart restaurant served something we called the Salad Sandwich. For customers still in the know, it's secretly available upon request and it's not a dainty choice. Just be prepared to open wide. This recipe, inspired by the *Food52* blog, is a different take on a salad in a sandwich. The bright marinated kale pairs so nicely with the buttery avocado and zest from our Pesto Vegenaise.

SERVES 2

1 bunch lacinato kale, stemmed and chiffonaded

2 tablespoons fresh lemon juice

Sea salt and freshly ground black pepper

1 small shallot, peeled and thinly sliced

2 small radishes, thinly sliced

¼ cup Follow Your Heart Shredded Parmesan

3 tablespoons olive oil

¼ cup Pesto Vegenaise

4 slices of your favorite vegan whole wheat sandwich bread

1 ripe avocado, pitted and peeled

Toss the shredded kale with 1 tablespoon of the lemon juice and a pinch of salt. Massage the lemon juice into the kale and set aside.

Meanwhile, toss the sliced shallot with the remaining lemon juice and set aside. Let both the kale and the shallot marinate for at least 30 minutes. Then, toss them together along with the sliced radishes, Parmesan, olive oil, and salt and black pepper to taste.

For each sandwich, spread the Pesto Vegenaise on one side of two slices of bread. Smash half an avocado on one side of each of the remaining two slices and sprinkle with a pinch of salt. Pile on some kale salad, close the sandwiches, and serve.

Pepper Jack Roasted Red Pepper and Avocado Tartine

In the early days of Follow Your Heart, we got our bread from a small outfit known as Mrs. Williams Bakery. This was the freshest, most luscious whole wheat bread I had ever tasted and it lent its special character to so many of our menu items. When Mrs. Williams closed, we sought to replicate that bread, and to this day, it stands as the benchmark for our sandwich bread that we now make in-house. When you combine fresh, farmers' market vegetables with your highest-quality locally or home-baked bread, you can just about wow every time. In this recipe, the sweetness of the red bell peppers and the acid of the tomatoes balance the rich avocado and everything is opened up with a little heat from our Pepper Jack Slices.

SERVES 2

2 large red bell peppers

¼ cup Vegenaise

2 slices vegan sourdough bread

4 Follow Your Heart Pepper Jack Slices

3 small heirloom tomatoes, thinly sliced

1 avocado, pitted, peeled and thinly sliced

Sea salt and freshly ground black pepper

Preheat the broiler and roast the peppers on a baking sheet for about 5 minutes per side, or until charred. Remove from the broiler, set aside, and allow to cool. Once cool enough to handle, peel away the skin and remove the stem, seeds, and membranes. Chop roughly and set aside.

Preheat the oven to 375°F.

Spread 2 tablespoons of Vegenaise on one side of each slice of bread. Top each slice with two slices of Pepper Jack and bake for about 10 minutes, or until the Pepper Jack is melted. Top with the tomato, roasted bell pepper, avocado, salt, and black pepper. Serve while hot and gooey.

Portobello French Dip with Savory au Jus

Two Los Angeles restaurants lay claim to the original French Dip, but we pride ourselves on the original vegan version. The earthy umami of the mushrooms creates such a savory and decadent 'jus' that everyone who tries this sandwich will swoon. We recently renewed this one as a lunch special and it has quickly become a local, vegan favorite. Important note: You will need a napkin for this one.

SERVES 2

2 large portobello mushrooms

2 tablespoons olive oil

1 medium onion, thinly sliced

3 garlic cloves, crushed and minced

1 cup low-sodium veggie broth

Juice of ½ orange (about 1 tablespoon)

1 tablespoon soy sauce

1 tablespoon vegan Worcestershire sauce

¼ teaspoon liquid smoke

½ teaspoon chopped fresh thyme

¼ teaspoon freshly ground black pepper

To Assemble

2 vegan French rolls

4 slices Follow Your Heart Provolone or Pepper Jack

Horseradish Sauce (page 148)

Clean the mushrooms with a damp paper towel, remove the stems, and scrape out the gills. Slice the caps into thin strips.

Heat 1 tablespoon of the olive oil in a large skillet over medium heat. Add the onion and cook until soft and translucent, about 20 minutes. Add the garlic and cook until the onion is caramelized and browned, 5 to 7 more minutes. Remove from the skillet and set aside on a plate.

Increase heat to medium-high, heat 1½ teaspoons of the remaining olive oil and add about half of the portobellos. Cook, stirring and tossing, until browned, 5 to 7 minutes. Remove from the pan and repeat with the remaining ½ teaspoon of olive oil and remaining mushrooms.

Return all the mushrooms to the skillet and add the onion and garlic as well as the veggie broth, orange juice, soy sauce, Worcestershire, liquid smoke, thyme, and black pepper. Bring to a low boil, then lower the heat to a simmer. Cook until the liquid has reduced by half, about 7 minutes.

continued ▶

Remove the mushrooms and onions with a slotted spoon, reserving the "jus" for serving.

Preheat the broiler.

Cut the French rolls in half and melt two slices of Provolone or Pepper Jack on the bottom half of each roll in the broiler for 2 to 3 minutes. Add the portobello mixture and spread the Horseradish Sauce on the cut side of the top half of each roll. Close the sandwich and serve with the "jus" for dipping.

Farro and Mushroom Sliders with Garlic Aïoli

These sliders are inspired by our Nutburger, which was created by Taryn, another of our talented cooks, in the late '70s. The Nutburger is one of our longest surviving recipes on the menu and these sliders are an at-home take on this FYH favorite. When the weather starts to turn, we indulge in the smoky balance of our Smoked Gouda with the bright zest of Organic Garlic Aïoli Vegenaise, each lightly complementing the hearty texture and nutrient-dense mushroom and farro patties.

YIELDS 8 TO 10 SLIDERS

3 cups veggie broth

1 cup uncooked farro

3 tablespoons neutral-tasting oil (safflower or grapeseed)

1 small onion, diced

2 cups button or cremini mushrooms, chopped

2 garlic cloves, finely diced

¼ teaspoon freshly ground black pepper

1 tablespoon vegan Worcestershire sauce

1 tablespoon fresh lemon juice

¼ cup Vegenaise

1 cup bread crumbs

1 teaspoon sea salt

1 teaspoon dried tarragon

3 slices Follow Your Heart Smoked Gouda, quartered

8 to 10 vegan wholegrain slider buns

¼ cup Organic Garlic Aïoli Vegenaise or the Basic Garlic Aïoli (page 147)

¼ cup Dijon mustard

2 cups fresh arugula

1 tomato, sliced

Bring the veggie broth to a boil in a medium saucepan. Add the farro, lower the heat to medium, and cook with the lid askew for 20 to 25 minutes, until tender. Drain the cooked farro and transfer to a food processor.

Heat 2 tablespoons of the oil in a large skillet over medium-high heat and sauté the onion for 5 minutes, stirring occasionally. Lower the heat to medium. Add the mushrooms and cook for an additional 10 minutes, or until beginning to caramelize. Add the garlic and cook for 1 minute more.

Add the black pepper, Worcestershire sauce, lemon juice, mushroom mixture, and Vegenaise to the farro in the food processor and pulse to fully incorporate, scraping the sides often. Add the bread crumbs, salt, and tarragon and pulse until well blended. Transfer to a large bowl.

Form into eight to ten patties and set aside on a parchment-lined baking sheet.

continued ▶

Heat a large skillet over medium heat and lightly coat with the remaining 1 tablespoon oil. Cook each patty for 4 to 6 minutes per side, until golden brown and cooked through. After flipping and in the last minute, add a slice of Smoked Gouda to each and cover for a nice and even melt.

Toast your buns and spread with the garlic aïoli and Dijon mustard. Build your sliders with the arugula, tomato, and patties.

Double-Decker Spicy Chipotle M-L-A-T Sandwich

I know a lot of vegetarians and vegans who aren't huge fans of meat alternatives. Whether the alternatives are too close for comfort or they just prefer something less processed, I've found mushrooms can stand in for a lot of different meats. These marinated and baked shiitakes, paired with our Chipotle Vegenaise and ripe tomato and avocado, complete the perfect spicy M-L-A-T.

SERVES 2

Mushroom "Bacon"

2 cups fresh shiitake mushrooms

¼ cup hickory liquid smoke

1 tablespoon olive oil

1 teaspoon sea salt

1 teaspoon freshly ground black pepper

1 teaspoon smoked paprika

To Assemble

1 cup Chipotle Vegenaise

6 slices vegan sourdough or classic white bread

4 leaves red or green leaf lettuce

8 slices ripe tomato

1 avocado, pitted, peeled, and sliced

Prepare the "bacon": Preheat the oven to 350°F.

Clean the mushrooms with a damp paper towel, remove the stems, and scrape out the gills. Slice the caps into ⅙- to ⅛-inch strips.

Marinate the mushrooms in the liquid smoke for about 5 minutes. Line a baking sheet with aluminum foil. Toss the mushrooms in the olive oil, salt, black pepper, and paprika, then arrange in a single layer on the prepared baking sheet. Roast in the oven for about 20 minutes. Using metal tongs, flip each piece of shiitake and roast for another 20 minutes, or until slightly crispy. Remove from the oven, drain on paper towels, and set aside.

Assemble the sandwiches: Spread the Chipotle Vegenaise on one side of each slice of bread. On each of two slices, lay one lettuce leaf, two slices of tomato, one-quarter of the avocado, and one-quarter of the mushroom "bacon." Cover each stack with a second slice of bread and repeat adding the lettuce, tomato, avocado, and mushroom "bacon." Close each sandwich and open wide.

Ultimate Grilled Veggie Panini

Every couple of years, we update our menu to reflect new options that cater to changing tastes. While we keep our core offerings, we venture into new forms; a few years ago, we added a set of three panini. Although they have since grown out of fashion again, here is a recipe you can make at home inspired by their brief life on our menu. Not only are these easy to make, the simplicity of the crunchy bread and the subtle, garlic-infused grilled vegetables make these panini a frequent lunch favorite.

SERVES 2

1 zucchini, sliced lengthwise into ¼-inch strips

1 summer squash, sliced lengthwise into ¼-inch strips

1 red onion, sliced into ¼-inch-thick rings

¼ cup Roasted Garlic Vegenaise

Salt and freshly ground black pepper

4 large slices vegan sourdough bread

1 roasted red bell pepper, sliced

4 slices Follow Your Heart Mozzarella

Heat a grill pan over medium heat.

Place the zucchini, squash, and onion in a medium bowl. Coat with 2 tablespoons of the Roasted Garlic Vegenaise and sprinkle with salt and black pepper to taste. Grill the vegetables until tender and grill marks appear, about 8 minutes, turning once. Remove from the pan and let cool.

Coat both sides of the bread slices with the remaining Roasted Garlic Vegenaise. Fill each sandwich with roasted red pepper, the grilled veggies, and two slices of the mozzarella. Close each sandwich and cook in a panini press for 3 to 4 minutes, or in a grill pan with another heavy pan on top, flipping once, until golden brown.

Gourmet Grilled Cheese Hack

This hack has gained mainstream attention in the last year but remains of one of our tried-and-true secrets in the Follow Your Heart kitchen. Simple, yet surprisingly effective at getting the perfect level of crisped and golden-browned slices of bread on the outside and an even melt of American Slices on the inside.

YIELD 2 SANDWICHES

4 large slices vegan sourdough or wholegrain bread

2 to 3 teaspoons Vegenaise

4 slices Follow Your Heart American Slices

Preheat a skillet or griddle over medium heat.

Lightly coat both sides of each bread slice with Vegenaise.

Place one slice of bread on the hot skillet and add two American Slices. Cover with a second piece of bread. Cook for about 2 minutes, or until golden brown. Carefully flip and cook for an additional minute, or until the slices are melted and the bread is golden brown. Repeat to cook the other sandwich.

Bob Says

You can get creative with the flavors of Vegenaise and cheese you choose for some interesting combinations. I like our Smoked Gouda with some Organic Garlic Aïoli Vegenaise and I throw in fresh arugula for color and vitamin A.

Crispy Mushroom Po' Boy Sandwiches

This recipe was brought to me by Katie, who proudly refers to herself as the "condiment queen." She found her way to our store 20 years ago, solely based on her love of Vegenaise, and a little bird that told her we were hiring. She started in the cashier department, and has climbed the ranks to becoming one of our VPs. A humble nod to Southern cuisine, these mushroom po'boys are crispy and full of Cajun spice, and exist if for no other reason than to serve as a vessel on which to deliver our Cajun Sauce (page 156).

YIELDS 4 SANDWICHES

1 pound cremini or button mushrooms

½ cup Vegenaise

¼ cup unsweetened almond milk

1 cup unbleached all-purpose flour

2 teaspoons sea salt

1 teaspoon cayenne pepper

1 teaspoon garlic powder

1 teaspoon onion powder

1 teaspoon freshly ground black pepper

½ teaspoon smoked paprika

2 cups neutral-tasting oil (safflower or grapeseed) for frying

4 vegan French rolls or hoagie buns

1 cup Cajun Sauce (page 156)

2 tomatoes, sliced

1 cup shredded green cabbage

Clean the mushrooms with a damp paper towel and remove the stems.

Place the Vegenaise in a small bowl and whisk in the almond milk until smooth.

Whisk together the flour, salt, cayenne, garlic powder, onion powder, black pepper, and paprika in a medium bowl until incorporated. Position this bowl next to the Vegenaise bowl.

Coat the mushrooms in the Vegenaise mixture, then switch hands and dredge in the flour mixture. Place the dredged mushrooms on a baking sheet and set aside.

Heat the oil in a large, heavy-bottomed pot or cast-iron skillet over medium-high heat. When the oil is hot (350°F), use a slotted metal spoon to fry the mushrooms in batches. Cook until golden brown and crunchy. Transfer to a paper towel–lined wire rack to cool slightly and absorb any excess oil.

Toast the French rolls and assemble the sandwiches with a generous smear of Cajun Sauce, tomatoes, cabbage, and crispy mushrooms. Serve hot.

SALADS

Spanish Chickpea Salad

This dish brings together many of my favorite ingredients, specifically chickpeas and cashews, and combines these excellent sources of protein in a new way. Paprika is an ingredient that you can pull from the spice cabinet when looking to make something with warmth—and not warm because of the oven or stove temperature. It just seems to brighten your soul and this chickpea salad has so many nuances of flavor and texture that satisfy on several levels.

SERVES 2 TO 4

Two 15-ounce cans chickpeas

1 cup Vegenaise

Zest of 1 lemon

1 tablespoon fresh lemon juice, plus more to taste

1 tablespoon smoked paprika

1 garlic clove, crushed and minced

1 teaspoon vegan Worcestershire sauce

1 teaspoon sea salt, plus more to taste

4 green onions, thinly sliced

¼ cup fresh parsley, chopped

1 teaspoon crushed red pepper flakes

1 cup toasted, salted cashews

Place the chickpeas in a large bowl filled with water. Gently massage the chickpeas to remove their skin. Drain and pat dry.

Whisk together the Vegenaise, zest and lemon juice, paprika, garlic, Worcestershire sauce, and salt in a medium bowl.

Add the chickpeas and mash with a fork or potato masher until about a little more than half are smashed; let some remain whole.

Gently stir in the green onions, parsley, red pepper flakes, and cashews. Add more salt and lemon juice to taste.

Let sit for at least 15 minutes before serving. Serve chilled or at room temperature and top with the toasted cashews.

Corn and Curry Potato Salad

Growing up in Illinois where corn is king, I gained a deep appreciation for sweet corn early on, sometimes taking it right off the plant and eating it raw while standing in the field, dwarfed beneath the towering stalks. In this recipe, the sweetness of the corn complements the earthy flavor of the curry and cumin and this spin on a classic potato salad may become a new summer classic.

SERVES 4 TO 6

8 small red potatoes, scrubbed and cut into quarters

1 tablespoon neutral-tasting oil (safflower or grapeseed)

1 cup Vegenaise

2 teaspoons curry powder

¼ teaspoon ground cumin

1 teaspoon sea salt

½ cup finely diced red onion

½ cup corn kernels

¼ cup finely chopped fresh parsley, plus more for garnish

¼ cup peanuts, chopped

Preheat the oven to 400°F.

Coat the potatoes in the oil and arrange evenly in a single layer on a baking sheet. Roast in the oven until easily pierced with a fork, 25 to 30 minutes (depending on potato size). Remove from the oven and let cool completely, then dice into large cubes.

Whisk together the Vegenaise, curry powder, cumin, and salt in a large bowl until uniformly yellow.

Add the potatoes, red onion, corn, parsley, and peanuts and gently fold until all the ingredients are evenly coated. Serve.

Beluga Lentil Salad with Minty Crème Fraîche

SERVES 2 TO 4

Many years ago, our family went on a trip to France with our daughter Claire, who today directs much of the menu development at Follow Your Heart. She fell in love with the simplicity of French ingredients and their delicate texture as exemplified in this fresh and light lentil salad. Mint surprisingly pairs nicely in Vegenaise and if you make the crème fraîche even a day in advance, you'll get more of that clean mint flavor to pair with the rich lentils and feta crumbles.

Lentils

1 shallot, finely chopped

¼ teaspoon ground cumin

1 tablespoon olive oil

1 garlic clove, crushed and minced

1 cup dried beluga lentils, rinsed

2 cups veggie broth

1 teaspoon sea salt

Minty Crème Fraîche

½ cup Vegenaise

¼ cup finely chopped fresh mint

1 teaspoon white wine vinegar

½ teaspoon sea salt

½ teaspoon freshly ground black pepper

To Assemble

1 cup packed arugula

1 cup toasted pine nuts

1 Persian or English cucumber, cubed

¼ cup Follow Your Heart Feta Crumbles

Leaves from 2 sprigs mint

Prepare the lentils: Sauté the shallot and cumin in the oil in a medium saucepan over medium-low heat until translucent. Add the garlic and cook for 1 minute. Add the lentils, broth, and salt and simmer over low heat, partially covered, for 15 to 20 minutes, until all the liquid is absorbed. Transfer the lentils to a baking sheet to cool.

Prepare the crème fraîche: Stir together the Vegenaise, mint, vinegar, salt, and black pepper in a small bowl and refrigerate for 30 minutes.

Assemble the salad: Serve the lentils on top of the arugula with a dollop of minty crème fraîche, pine nuts, cucumber, and Feta Crumbles. Garnish with fresh mint.

Pesto Pasta Salad

At Earth Island, where we make Vegenaise and our other Follow Your Heart products, we have a garden in the back for employees, where we harvest seasonal vegetables and herbs. Basil is by far the most abundant and one which we have to pick almost daily to keep up. We also had extremely productive cherry tomato plants that were over seven feet high this year. Luckily, they make the perfect pasta salad when combined. If you can't find our Pesto Vegenaise locally, simply puree some fresh basil in a food processor and fold into your Vegenaise of choice to taste.

SERVES 4 TO 6

1 pound dried small pasta of your choice (rotelle, rigatoni, shells, etc.)

One 10-ounce package frozen spinach, thawed, excess water pressed out

One 16-ounce bag frozen peas, thawed, rinsed, and drained

1½ cups Pesto Vegenaise

½ teaspoon salt

½ teaspoon freshly ground black pepper

1 pint cherry or grape tomatoes, halved

Cook the pasta according to the package instructions. Rinse with cold water, drain, and let cool to room temperature.

Mix together the spinach, peas, Pesto Vegenaise, salt, and black pepper in a large bowl. Add the pasta and halved tomatoes. Serve immediately or refrigerate for up to a day.

Simple Potato Salad

I love a good potato salad and, for me, it's not a good one unless it's made with Vegenaise. A good potato salad requires you to cook those potatoes just right so they're firm enough to keep their shape, but soft enough to break down just enough to hold it all together. We've made a lot of potato salad since we first started using Vegenaise in the restaurant. Here is one of our most simple and classic recipes that serves as the perfect picnic side dish because it doesn't contain eggs and won't spoil on the way.

SERVES 4 TO 6

1½ pounds red or white potatoes, peeled

1½ cups Vegenaise

1 tablespoon cider vinegar

1 tablespoon prepared yellow mustard

1 teaspoon sea salt

¼ teaspoon freshly ground black pepper

¼ cup Follow Your Heart Smoky Bits

1 medium red onion, chopped

Chopped fresh parsley for garnish

Paprika for garnish

Place the potatoes in a large pot and add enough water to cover. Cover and bring to a boil. Lower the heat to low and cook, uncovered, for 25 to 30 minutes, or until fork-tender. Drain and let cool. Cut the potatoes into cubes.

Stir together the Vegenaise, vinegar, mustard, salt, and black pepper in a large bowl.

Add the potatoes, Smoky Bits, and red onion and toss. Sprinkle with the parsley and paprika and refrigerate for at least 2 hours. Serve.

Spicy Southwestern Potato Salad

When we launched our gourmet Vegenaise line in 2009, we played with a core set of recipes to bring a new spin to our classic collection. Beginning with Original Vegenaise, we brought in new flavors and variations and worked with them until we felt they were ready for prime time. Only then did we introduce them to the market. This potato salad packs some punch by using our Chipotle Vegenaise, and the sweetness of the corn keeps you coming back for another bite.

SERVES 4 TO 6

2 pounds red potatoes

½ cup Chipotle Vegenaise

2 tablespoons chopped fresh cilantro

2 green onions, chopped

¾ cup corn kernels

2 tablespoons fresh lime juice

¼ teaspoon sea salt, or to taste

Place the potatoes in a large pot and add enough water to cover. Cover and bring to a boil. Lower the heat to low and cook, uncovered, for 25 to 30 minutes, or until fork-tender. Drain and let cool. Cut the potatoes into ¾- to 1-inch cubes.

Place the potatoes in a large bowl and mix gently with all other ingredients, adding salt to taste. Cover and store refrigerated until ready to serve.

Traditional Vegan Coleslaw

This classic slaw is not only a barbecue side dish requirement, but often becomes a topping for sliders, baked potatoes, or tacos. After making coleslaw in the restaurant for years, we finally launched a retail version of the coleslaw dressing, which was introduced just over a year ago. If a store near you carries the coleslaw dressing, save yourself some time, as we've done most of the work for you.

SERVES 2 TO 4

2 green onions, finely chopped

⅔ cup Vegenaise

1 teaspoon freshly ground black pepper, plus more to taste

1 teaspoon celery salt

3 tablespoons evaporated cane sugar

2 tablespoons cider vinegar

One 1-pound package shredded cabbage mix

Sea salt

Mix the green onions, Vegenaise, black pepper, celery salt, sugar, and vinegar in a large bowl to make the dressing. Stir in the shredded cabbage and toss to coat. Add salt and additional black pepper to taste.

Waldorf Salad

In the '70s and '80s, Waldorf salads were on just about every menu across the country, though they seem to have gone out of fashion in recent years. I can't think about this salad without it bringing to mind the *Fawlty Towers* sketch with John Cleese repeating over and over, "Celery, apples, walnuts, grapes" ... in a mayonnaise sauce. Of course, we prefer ours with Vegenaise. When you're invited to a potluck brunch and want to bring something that whips up quick but looks amazing, you really can't get any simpler than this old standby.

SERVES 2 TO 4

2 celery stalks, sliced

1 cup walnuts, crushed into pieces

2 apples (Fuji, Pink Lady, or Granny Smith), cored and diced

1 cup raisins

3 tablespoons Vegenaise

1 tablespoon fresh lemon juice

Sea salt and freshly ground black pepper

Mix together all the ingredients in a bowl, adding salt and black pepper to taste, and evenly coat.

Watermelon Radish, Romanesco, and Blood Orange Salad

When I was five years old, I had my first-ever garden and grew my first vegetable: a little red radish. While they've never become a particular favorite of mine, I do love a thinly sliced watermelon radish to add crunch, and they're absolutely gorgeous. This salad was inspired by a (sadly) short-lived restaurant here in Los Angeles called P.Y.T. With such strikingly beautiful farmers' market vegetables paired with the citrus balance of blood oranges, we let the vegetables shine with a hint of kick from the habanero dressing.

SERVES 2 TO 4

1 large head romanesco broccoli, quartered

1 large watermelon radish

¾ cup Vegenaise

Juice of 1 large lemon (about 2 tablespoons)

1 clove garlic, crushed and minced

1 small habanero pepper, seeded and thinly sliced

½ teaspoon ground cumin

¼ teaspoon sea salt

¼ teaspoon freshly ground black pepper, plus more to taste

¼ cup roughly chopped fresh cilantro

1 blood orange, peeled, pith removed, segments separated and cut in half

Edible flowers for garnish (optional)

Gently steam the romanesco for about 2 minutes, then allow to cool in a bowl of cold water.

Slice the watermelon radish as thin as you can, ideally with a mandoline.

Once the romanesco is cool, slice lengthwise, using a mandoline. You want very thin, pliable strips.

Prepare the dressing by whisking together the Vegenaise, lemon juice, garlic, habanero, cumin, salt, and black pepper in a small bowl. Taste and adjust the seasonings as needed. Add more Vegenaise if too spicy.

Combine the romanesco, radish, cilantro, and dressing in a large bowl and toss together. Add the blood orange pieces and gently mix.

Transfer the salad to a serving plate, top with another few cracks of black pepper, arrange the flowers (if using), and serve!

Asian Broccoli Slaw

Over the years, many fine chefs have stepped foot in our kitchen and cooked up some memorable meals. We've even had a few famous guest chefs, such as Tal Ronnen of Crossroads Kitchen, grace our stovetop and it was upon the launch of his first cookbook, *The Conscious Cook*, when we were able to bring him in for an evening treat and book signing. While what we prepare is nothing like what he served that night, and humbly far from his mastery of ingredients and flavors, he inspires us daily and we're regular patrons of Crossroads. An inspired fancier take on a traditional recipe that we've served for years is this broccoli slaw, which came as a side, back in 2016, when we served some Teriyaki Follow Your Heart Burgers with Grilled Pineapple and Soy Sauce Glaze.

SERVES 4 TO 6

Dressing

¼ cup Vegenaise

2 teaspoons sesame oil

2 tablespoons liquid aminos

3 tablespoons rice vinegar

1 tablespoon agave nectar

1 tablespoon finely chopped fresh ginger

2 tablespoons sesame seeds

Slaw

One 12-ounce bag broccoli slaw

1 carrot, finely shredded

2 tablespoons chopped green onion

1 cup mandarin orange slices, pith removed

2 tablespoons toasted slivered almonds

Prepare the dressing: Combine the Vegenaise, sesame oil, liquid aminos, rice vinegar, agave, ginger, and sesame seeds in a large bowl. Whisk until smooth.

Prepare the slaw: Add the broccoli and carrot shreds, green onion, and mandarin oranges to the dressing and sprinkle with the toasted slivered almonds.

Sriracha Corn Salad

There's no denying it, our Sriracha Vegenaise has some kick. A lot of kick. When we were developing this one, we considered the risk of making it too spicy, but at the end of the day, it had to be bold and live up to its name. Feel free to cut the heat with as much Original Vegenaise as you need, or like some of our chefs, just go full stop with all Sriracha, all the time.

SERVES 2 TO 4

2 tablespoons vegan butter

2 tablespoons olive oil

6 ears of corn, kernels removed from cob

2 medium red bell peppers, seeded and finely chopped

½ cup Sriracha Vegenaise

¼ cup Vegenaise

¼ cup Follow Your Heart Grated Parmesan

½ cup chopped fresh cilantro

Juice of 1 lime

Sea salt and freshly ground black pepper

Heat the vegan butter and olive oil in a large skillet over medium-high heat. Add the corn kernels and cook for 15 to 20 minutes, until the corn begins to brown slightly.

Add the bell peppers and cook for additional 6 to 8 minutes, stirring often.

Add ¼ cup of the Sriracha Vegenaise and stir to coat. Cook for an additional 2 to 4 minutes.

Remove from the heat and transfer the corn mixture to a bowl.

Fold in the remaining ¼ cup of Sriracha Vegenaise and the Vegenaise, Parmesan, cilantro, and lime juice and stir to coat completely.

Add salt and black pepper to taste and serve!

Summer Macaroni Salad

In the Follow Your Heart kitchen, we make many types of macaroni salad, some of which we sell in the deli section of our market. We've got a classic one, a smoked chick'n one, and here's one that's a great option to make at home featuring your fresh garden or farmers' market vegetables. This particular macaroni salad is a favorite because of the gorgeous colors of the red tomatoes, purple red onion, and green zucchini.

SERVES 4 TO 6

1 pound dried elbow macaroni

½ cup Vegenaise

2½ tablespoons red wine vinegar

2 teaspoons Dijon mustard

¼ teaspoon dried oregano

¼ teaspoon granulated garlic

½ cup chopped red onion

4 cups grape tomatoes, cut in half

1 small zucchini, thinly sliced thin and quartered

Sea salt and freshly ground black pepper

Cook the pasta in a large pot of salted water according to the package directions. Drain and rinse with cold water.

Stir together the Vegenaise, vinegar, mustard, oregano, garlic, and red onion in a large bowl. Gently add the tomatoes, zucchini, and pasta. Add salt and black pepper to taste.

Chickpea "Tuna" Salad Melt

Tuna salad is a dish from childhood that I thought I'd never again experience when I decided to go vegetarian. When our R&D scientist Jessica brought me this chickpea "tuna" recipe many years ago, I was transported back to boyhood. Top with vegan American Slices and this melt is simply decadent.

SERVES 2

¼ cup Vegenaise

2 celery stalks, finely chopped

2 green onions, finely chopped

¼ red bell pepper, seeded and diced

3 tablespoons prepared dill relish, or 1 dill pickle, chopped

1 tablespoon fresh lemon juice

1 garlic clove, mashed

1 teaspoon prepared mustard

½ teaspoon dried dill

1 tablespoon capers, drained

One 15-ounce can chickpeas, drained and rinsed

2 slices vegan sourdough bread

4 slices Follow Your Heart American Slices

Stir together all the ingredients, except the chickpeas, bread, and American Slices, in a large bowl, then fold in the chickpeas. Mash but do not puree. You want to leave a few small chunks for texture. Refrigerate for 30 minutes before serving.

Set your oven or a toaster oven to BROIL. Place a heaping spoonful of the "tuna" salad on a slice of bread and cover with two American Slices. Broil for 3 to 5 minutes, until the slices have melted. Repeat to make the other "tuna" melt.

Crudités with Miso Dip

Even though we have a robust organic produce section in the Follow Your Heart Market, we still support our excellent, local farmers' markets, not just for the additional variety, but also just because. Long seasons keep us in an abundance of beautiful, fresh fruits and vegetables. This recipe needs little else than your favorite vegetables to dunk with. Prepare a colorful platter of fresh market vegetables and this Miso Dip will pair nicely with every single one.

SERVES 4 TO 6

Assorted Vegetables

1 pound heirloom carrots, scrubbed and trimmed

8 radishes, scrubbed, trimmed, and sliced into wedges

2 large Belgian endives, trimmed and leaves separated

1 large head cauliflower, cut into florets

1 head radicchio, leaves removed and halved

1 pound sugar snap peas, fibers removed

6 Persian cucumbers, halved lengthwise or cut into spears

Miso Dip

1½ cups Vegenaise

¼ cup white miso paste

1 teaspoon fresh lime juice

1½ tablespoons rice vinegar

½ teaspoon finely chopped chives for garnish

Prepare the vegetables: Arrange the vegetables however you like on a serving platter or table.

Prepare the Miso Dip: Mix together the Vegenaise and white miso paste in a medium bowl. Just before serving, stir the lime juice and rice vinegar into the Miso Dip and top with the chives.

Serve with a few ice cubes over veggies, to keep them cool and fresh.

MAINS

Fried Ensenada Tacos and Mango Salsa

With our offices located in Southern California, tacos are not hard to come by and are arguably some of the best in the world. They inspired us to create what we've dubbed our "wish" tacos, which were quickly added to our menu a few years ago. In our kitchen, we fry tofu, but these tacos are an alternate at-home version and I love the sweetness of this Mango Salsa and the added crunch of the hearts of palm "fish" sticks.

SERVES 4 TO 6

Mango Salsa

2 teaspoons sea salt

¼ cup evaporated cane sugar

¼ cup white vinegar

1 small red onion, thinly sliced

1 mango, pitted, peeled, and chopped

¼ cup chopped fresh cilantro

Taco Filling

Two 14-ounce cans hearts of palm

1 cup unbleached all-purpose flour

½ cup Vegenaise

2 tablespoons fresh lime juice

2 cups panko bread crumbs

Neutral-tasting oil (safflower or grapeseed) for frying

Sea salt

To Assemble

Twelve 4.5-inch corn tortillas

3 tablespoons Chipotle Vegenaise

1 cup shredded green cabbage

Fresh cilantro leaves for garnish

Squeeze of fresh lime juice

Prepare the Mango Salsa: First, pickle the red onion by combining the salt, sugar, and vinegar in a small bowl. Add the onion and fully immerse. Marinate for 30 minutes. Add the mango and cilantro and stir until evenly incorporated. Set aside in the fridge.

Rinse the hearts of palm and cut in half lengthwise. Pat dry.

Prepare three bowls, one with the flour, one with the Vegenaise and lime juice whisked together, and the third with the bread crumbs. First, dredge the hearts of palm in the flour, then in the Vegenaise mixture, then in the bread crumbs, coating evenly at each bowl. Place on a parchment-lined baking sheet.

Heat the oil in a large, heavy-bottomed pot or cast-iron skillet over medium-high heat. When the oil is hot (350°F), use a slotted metal spoon to fry the hearts of palm in batches. Cook until golden brown and crunchy, about 2 minutes. Transfer to a paper towel–lined plate to cool slightly and absorb any excess oil. Sprinkle lightly with salt.

Assemble the tacos: Warm the tortillas over an open flame or in a dry skillet until hot. Stack two tortillas for each taco, then spread with 1½ teaspoons of Chipotle Vegenaise. Add two or three hearts of palm, a small handful of cabbage, and the mango salsa, and garnish with cilantro leaves and a squeeze of lime juice.

Vegan Cauliflower al Pastor Tacos

I. Love. Cauliflower. It's a favorite stand-in for meat and one that I can't help ordering every time I see it on a menu. Here, we have incorporated traditional al pastor flavors to make something which is a little bit sweet and a little bit spicy. Grilled pineapple and chipotle cauliflower really pair well in this version of cauliflower al pastor.

YIELDS 4 TACOS

Cauliflower al Pastor

1 large head cauliflower

3 garlic cloves, smashed

1 medium yellow onion, thinly sliced

½ cup Chipotle Vegenaise

2 tablespoons pineapple juice

2 tablespoons fresh lime juice

Freshly ground black pepper

2 tablespoons olive oil

2 teaspoons kosher or sea salt

To Assemble

3 pineapple rings

Four 4.5-inch flour tortillas

Thinly sliced radish

Fresh cilantro

Salsa verde for serving

Prepare the cauliflower: Stem and cut the cauliflower into florets. Combine the cauliflower with the garlic, onion, Chipotle Vegenaise, pineapple juice, lime juice, and black pepper to taste in a shallow bowl. Massage until well mixed and evenly coated. Cover and let marinate in refrigerator for at least an hour, overnight preferred.

Heat the oil on a grill pan over medium-high heat. Add the cauliflower mixture and salt and cook, covered, until browned and the onion is caramelized, about 15 minutes, stirring occasionally to ensure nothing sticks to the bottom of the pan. Once the cauliflower is lightly charred, remove it from the pan and set aside.

Assemble the tacos: Place the pineapple rings in the grill pan and cook until lightly charred and caramelized, 2 to 3 minutes per side. Remove from the pan and roughly chop.

Heat the tortillas in the grill pan until warm, about 1 minute.

Serve the cauliflower in the tortillas topped with the pineapple chunks, radish, and cilantro. Serve with salsa verde.

Cheesy Chipotle Potato Flautas

An office party favorite, these flautas are wonderfully cheesy, crispy, spicy, and satisfying. Finding vegetarian flautas in anything but a vegetarian or vegan restaurant is not easy, as they customarily contain meat and are almost never made to order. One of our favorite vegan restaurants that has vegan flautas inspired this recipe. The kick of Chipotle Vegenaise is not one to mess with if you're shy around spice . . . just meditate through the burn!

SERVES 4 TO 6

1½ pounds russet potatoes, scrubbed and peeled

¼ cup Chipotle Vegenaise

1 teaspoon ground cumin

1 teaspoon smoked paprika

1 teaspoon chili powder

1 tablespoon sea salt

1 tablespoon freshly ground black pepper

½ cup Follow Your Heart Fiesta Blend Shreds

6 large flour tortillas

1 cup neutral-tasting oil (safflower or grapeseed)

Follow Your Heart Dairy-Free Sour Cream for serving

Pico de gallo for serving

Guacamole for serving

Bring salted water to a boil in a large, heavy-bottomed pot. Add the potatoes and boil until fork-tender, 20 to 30 minutes. Drain the potatoes and return them to the pot.

Increase the heat to medium-high and add the Chipotle Vegenaise, cumin, paprika, chili powder, salt, and black pepper. Using a fork or potato masher, mash and mix the potatoes. Add the Fiesta Blend Shreds, mix well, and cook for 15 to 20 minutes, until the shreds have melted.

Place the tortillas in a lightly dampened paper towel, heat them for 30 seconds per tortilla. Spread an even amount of potato mixture on one end of each tortilla and roll up. Pierce with a toothpick to keep the flauta closed.

Heat the oil in a large, heavy skillet. Fry the flautas in batches of three for 2 minutes per side. Drain on a paper-towel-lined wire rack.

To serve, slice each flauta on a bias, top with the sour cream, and serve with pico de gallo and guacamole.

Green Garlic Broth with Slathered "Cheesy" Toast

Here, in the cold Southern California winters (hey, it's all relative), I've been known to sip steaming broth from large mugs at my desk and at home. This recipe turns this daily sip into a full meal. Miso-based broth, specifically live-fermented miso broth, has probiotic properties and is part of what keeps me healthy. Showcasing elegant and modest flavors, this combination of the garlic broth and the pine nut "cheesy" toast will warm you on even the coldest days.

YIELDS 2 TOASTS

Green Garlic Broth

2 garlic heads

3 tablespoons miso

4 cups filtered water

1 bunch green chard

"Cheesy" Spread

½ cup pine nuts, plus more for garnish

1 tablespoon miso

2 tablespoons Vegenaise

2 tablespoons water

To Assemble

2 slices multigrain bread

1 teaspoon everything bagel seasoning

Preheat the oven to 375°F.

Begin the Green Garlic Broth: Remove the outer paper layers of the garlic heads. Cut the tops off by about ¼ inch, exposing the cloves, wrap in foil, cut side up, and roast in the oven for 30 minutes, or until tender. Remove from the oven and set aside to cool.

Prepare the "Cheesy" Spread: Soak the pine nuts in hot water for 5 minutes. Drain and place in a blender along with 1 tablespoon of the miso, the Vegenaise, and 2 tablespoons of fresh water. Blend until smooth, scraping down the sides as needed. Remove from the blender and set aside.

Finish preparing the broth: Once the garlic has cooled, squeeze out the cloves and place them in the rinsed blender along with the 3 tablespoons of miso and the filtered water. Blend until smooth.

Transfer the broth mixture to a medium pot and bring to a simmer over medium-high heat. Turn off heat.

Stem the chard, chop into roughly 1-inch pieces, and add to the broth. Allow to sit for 1 minute, then ladle the broth into two bowls.

Assemble the toasts: Toast the bread and top with the "cheesy" spread, extra pine nuts, and the bagel seasoning. Serve with the Green Garlic Broth.

Cheesy Sweet Potato and Black Bean Sopes

SERVES 4 TO 6

Sopes are essentially small, fat tortillas with a tartlike edge that holds in whatever fillings your heart desires. Back in the '70s, Paul and I spent a lot of time traveling in Mexico, where we were first introduced to sopes, and we brought the concept back with us for a few dinners over the years. This recipe incorporates mashed sweet potatoes combined with black beans and Follow Your Heart Fiesta Blend Shreds for a complex, sweet, and spicy base topped with crunchy veggies, sweet pineapple, and smoky Chipotle Vegenaise.

Filling

1 pound sweet potato (2 medium or 1 large), scrubbed

2 tablespoons neutral-tasting oil (safflower or grapeseed)

½ medium yellow onion, diced

1 jalapeño pepper, seeded and diced

1 garlic clove, crushed and minced

½ teaspoon ground cumin

Pinch of sea salt, or to taste

One 15-ounce can black beans, drained and rinsed

1 cup Follow Your Heart Fiesta Blend Shreds

Sopes

1½ cups Mexican masa harina

½ teaspoon sea salt

1¼ cups warm water

Neutral-tasting oil (safflower or grapeseed) for pan (optional) and brushing

Toppings

½ cup shredded red cabbage

2 tablespoons roughly chopped fresh cilantro

1 small radish, thinly sliced

¼ cup pineapple chunks

½ cup Chipotle Vegenaise

Prepare the filling: Preheat the oven to 400°F.

Pierce the sweet potato several times with a fork and tightly wrap in foil. Bake for 45 to 60 minutes, until soft. (Alternatively, you can steam, microwave, or boil the sweet potato—whatever method you prefer to get it soft and mashable.) When the potato is cool enough to handle, remove and discard the skin and place the flesh in a medium bowl. Mash with a fork or potato masher until smooth.

Heat the oil in a large skillet over medium-high heat, then sauté the onion, jalapeño, and garlic until the onion and jalapeño are soft and translucent, 5 to 7 minutes. Add the cumin and salt, then add the mashed sweet potato, beans, and Fiesta Blend Shreds. Cook, stirring frequently, for 2 to 3 minutes, until the shreds have melted and the beans are well incorporated into the sweet potato. Add a splash or two of water if the mixture is too dry.

continued ▶

Prepare the sopes: Whisk together the masa harina and the salt in a medium bowl to evenly distribute. Then, add the warm water and knead until a firm but malleable dough forms. Add a teaspoon or two of water if too dry, or a teaspoon or two of masa if too wet. Divide the dough into eight equal portions. Cover the bowl with a clean, damp kitchen towel to keep them from drying out.

Preheat a large skillet over medium heat and lower your oven temperature to 375°F. Oil a baking sheet or line with parchment paper and set aside.

Roll each portion of the dough into a ball and then flatten in your hands into a patty about ¼ inch thick and 3½ inches in diameter. Pinch the sides of each sope to create a ridge about ¼ inch above the top of the patty, like the edge of a tart.

When the skillet is heated, place a few sopes at a time, flat side down, on the hot surface and let cook for 2 to 3 minutes, until the dough has become somewhat firm (but not dried out) and the bottom has a few browned spots. Then, place them on the prepared baking sheet. Brush the sopes with oil, especially the ridges and sides, and bake for 12 to 14 minutes, until firm. Keep warm at 250°F until ready to serve, or remove the pan from the oven and cover with foil to keep them moist.

Assemble the sopes: Fill each sope with a generous portion of the sweet potato mixture (you may end up with some filling left over). Pile high with the toppings, including a hearty dollop of Chipotle Vegenaise. Serve with plenty of napkins!

Classic Lasagna

Using tofu as ricotta may not be a new thing anymore, but we've been cooking up this lasagna for over 20 years. Well-drained tofu is key and Vegenaise keeps it textured as "ricotta" even after baking. If Kathy had her way, we'd exclusively use our Organic Garlic Aïoli Vegenaise, which I agree is arguably better, though less available across the United States. She also has a tendency to add our Garlic Aïoli to dishes that just need that extra-creamy, silky texture, such as soups or casseroles. Don't be afraid to get your hands dirty when you mash the ricotta. Feel free to riff on this recipe, but know that classic is always delicious.

SERVES 8 TO 10

Tofu Ricotta

Two 14-ounce containers firm tofu, well drained and patted dry with a paper towel

1 cup Vegenaise

1 teaspoon garlic powder

2 tablespoons Italian seasoning

1 tablespoon sea salt, or more to taste

1 teaspoon freshly ground black pepper

One 8-ounce package Follow Your Heart Mozzarella Shreds

Lasagna

1½ (16-ounce) boxes lasagna noodles (12 noodles)

1 teaspoon neutral-tasting oil (safflower or grapeseed) for greasing

Two 24-ounce jars vegan tomato-basil pasta sauce

Two 10-ounce packages frozen chopped spinach, thawed and well drained (squeeze out all the water)

½ cup Follow Your Heart Mozzarella Shreds

½ cup Follow Your Heart Shredded Parmesan

Prepare the tofu ricotta: Combine the tofu, Vegenaise, spices, salt, black pepper, and mozzarella in a large bowl. Using your hands or a potato masher, mash together until smooth and creamy, then refrigerate for at least 30 minutes.

Prepare the lasagna: Bring a large pot of salted water to a boil. Cook the lasagna noodles, a few at a time, until they are pliable but not overcooked, 4 to 6 minutes or per the package directions. Lay the cooked noodles flat on a lightly greased baking sheet to cool.

Preheat the oven to 350°F.

Spread an even layer of the sauce about ⅛ inch thick in a deep 9-by-13-inch baking dish. Arrange a layer of lasagna noodles to completely cover the sauce. Evenly spread one-third of the tofu ricotta over the pasta. Arrange another layer of lasagna noodles on top. Spread an even layer of sauce on the pasta, then half of the spinach, then another layer of one-third of the tofu ricotta. Arrange another layer of lasagna noodles over the ricotta, evenly spread with the remainder of the spinach and then the remaining tofu ricotta. Add one last layer of noodles and top with the remaining sauce. Sprinkle the ½ cup of mozzarella on top and top with a sprinkle of shredded Parmesan. Cover with foil and place on a baking sheet.

continued ▶

Bake for 45 minutes. Remove the foil, then broil for 5 minutes, or until the mozzarella and Parmesan are bubbly and golden brown. Remove from the oven and let cool at least 10 minutes before serving.

Bean Cutlets with Harissa Aïoli

I believe that vegans and vegetarians are always looking for something to make into a patty, to replace the many types of patties that we can't eat. I love chickpeas and while they're mostly known as the main ingredient for hummus, in this recipe, we blend them with black-eyed peas, mash them up, and make wonderful, high-protein patties out of them.

YIELDS 8 CUTLETS

¼ cup olive oil

6 large green onions, finely chopped

2 teaspoons sea salt

4 large garlic cloves, pressed or minced

One 15-ounce can chickpeas, drained and rinsed

One 15-ounce can black-eyed peas, drained and rinsed

3 tablespoons prepared mustard

2 teaspoons paprika

2 tablespoons Vegenaise

Harissa Aïoli (page 171) for serving

Heat 2 tablespoons of the olive oil in a medium skillet over medium-low heat. Add the green onions and salt and sauté until sweated and fragrant, about 10 minutes. Add the garlic and cook for 2 minutes more.

Place the chickpeas in a large bowl filled with water. Gently massage the chickpeas to remove their skin. Drain and pat dry.

Place the chickpeas and black-eyed peas in a bowl and mash with a fork or potato masher until only a few visible beans remain. Add the mustard, paprika, green onion mixture, and Vegenaise, mixing thoroughly. Refrigerate the mixture for at least 30 minutes.

Hand form eight to ten round cutlets, each about 1 inch thick.

Heat the remaining 2 tablespoons of olive oil in a large skillet over medium heat. Fry each cutlet for about 4 minutes each side, or until nicely browned but not burned.

Serve the cutlets with the Harissa Aïoli.

Grilled Marinated Tofu with Spicy Sour Cream Adobo

This delectable dish was created with my eldest son, David, in mind as he loves his chile peppers. Tofu absorbs marinade far better than meat as it's more porous; this has been a trick in our kitchen for years. The grilled tofu is simple, while the salsa and sour cream adobo sauces really steal the show.

SERVES 2

Spicy Sour Cream Adobo Sauce

2 canned chipotles in adobo sauce

½ cup Follow Your Heart Dairy-Free Sour Cream

Pinch of salt

1 teaspoon fresh lime juice

Roasted Corn Salsa

2 tablespoons neutral-tasting oil (safflower or grapeseed)

1 ear of fresh corn, husked

2 to 3 heirloom tomatoes, seeded and diced

1 tablespoon chopped fresh cilantro

Sea salt

Juice of 1 lime

Tofu

Juice of 1 lime

¼ cup Vegenaise

½ cup olive oil

Sea salt

1 teaspoon agave nectar

1 garlic clove, chopped

One 14-ounce container firm tofu

½ avocado, pitted, peeled, and sliced

Prepare the adobo: Place one to two of the canned chipotles in adobo sauce in a blender and blend until smooth. Transfer 1 tablespoon of the blended chipotle mixture to a small bowl and whisk with the sour cream, salt, and lime juice until smooth. Taste and add additional chipotle based on your desired heat level. Set the sauce aside in the refrigerator.

Prepare the salsa: Heat the oil in a medium skillet. Cut the corn off the cob, then blister the kernels in the hot oil until they have a nice brown color, 2 to 5 minutes. Transfer the corn to a medium bowl and set aside to cool.

Once the corn is cool, add the tomatoes and cilantro. Add sea salt to taste and a generous squeeze of lime juice. Toss the salsa gently, taking care not to crush the tomatoes.

Prepare the tofu: Create a marinade for the tofu by whisking together the lime juice, Vegenaise, olive oil, pinch of salt, agave nectar, and garlic in a medium bowl until blended.

Slice the tofu lengthwise, creating flat ¼-inch slabs, then soak in the marinade for 30 minutes.

Heat a grill or grill pan and place the marinated tofu on the grill. Allow grill marks to develop and then flip once, cooking for 2 to 3 minutes per side. When the tofu is done cooking, place it in a shallow dish and pour a generous portion of the marinade over it, coating each side.

To assemble: With a slotted spoon, place a generous portion of corn salsa on each plate. Balance a piece of tofu on top. Garnish the top with sliced avocado. Drizzle the adobo over the top.

Arugula White Pizza

This recipe came to us at a tradeshow when our friend, Chef Proof, showcased both of our focus items in one addicting pizza. Who needs pizza sauce when you have garlic aïoli? It was there in London, where two plant-based universes collided, that Kathy and I attended a tempeh class, hosted by Seth Tibbott of Tofurky, who also happens to be a tempeh master. In the class, we made some amazing loafs of fresh tempeh that he took back to his hotel, where he had the perfect fermentation environment for the tempeh to develop. Two days later, he brought the tempeh to the show with him. In a moment of inspiration, Seth came over to the Follow Your Heart booth where he and I proceeded to sauté the tempeh pieces and Chef Proof added them to the pizzas he was serving, along with our Garlic Aïoli. Seth and I had a blast huddled over a tiny skillet with two little spatulas and we cooked up every piece of tempeh we had made and even more that he brought with him. Simple ingredients are simply delicious, and good times will be had by all.

YIELDS 1 PIZZA

One 12-inch unbaked vegan pizza crust

All-purpose flour for dusting

2 cups fresh arugula

1½ tablespoons olive oil

1 tablespoon fresh lemon juice

½ cup Garlic Aïoli Vegenaise

1 tablespoon Italian seasoning

½ cup Follow Your Heart Shredded Parmesan

One 8-ounce package tempeh, sliced into patties (optional)

1 tablespoon neutral-tasting oil (safflower or grapeseed) for searing (optional)

Preheat the oven to 400°F.

Roll out the pizza dough on a floured surface. Transfer to a baking sheet.

Mix together the arugula, olive oil, and lemon juice in a medium bowl and let marinate for 5 minutes.

If adding tempeh, heat oil in a large skillet over medium-high heat and sear tempeh until light brown, flipping once, 2 to 3 minutes per side.

Spread the Garlic Aïoli Vegenaise over the dough and sprinkle with the Italian seasoning and shredded Parmesan. Add the marinated arugula, allowing any excess liquid to drain back into bowl. Bake for about 20 minutes, or until the dough is starting to brown. Remove from the oven, allow to cool for about 3 minutes, and serve!

Pesto Artichoke Pizza

Since pesto usually contains Parmesan, it's often not an option for vegans, leading us to create our own version for use in the restaurant. Back in the early '90s, we actually made a Pesto Dressing that was out of this world, and when its day came up on the production schedule, our entire block smelled of that delicious fresh basil while it blended. Years later, we introduced Pesto Vegenaise and we once again get to enjoy that wonderful aroma. Pesto Vegenaise in place of traditional tomato sauce adds a freshness that, when paired with the artichokes and arugula, really brings the garden to the table.

SERVES 4 TO 6

2 tablespoons Pesto Vegenaise

One 12-inch unbaked vegan pizza crust

1 cup Follow Your Heart Mozzarella or Pizzeria Shreds

2 cups marinated artichoke hearts, roughly chopped

1½ cups fresh arugula leaves

1½ tablespoons fresh lemon juice

Preheat the oven to 450°F with a rack on the lowest level.

Roll out the pizza dough on a floured surface. Transfer to a baking sheet.

Spread the Pesto Vegenaise evenly over the pizza crust and top with ½ cup of the mozzarella. Arrange the artichokes evenly over the pizza and sprinkle with the remaining mozzarella. Bake for 11 to 12 minutes, until the mozzarella is fully melted.

While the pizza is baking, place the arugula in a bowl, add the lemon juice, and toss to combine.

Top the pizza with the arugula, slice, and serve!

Sautéed Mushroom Chipotle Quesadillas

For over 20 years, we've served a very filling and popular grilled vegetable quesadilla in the appetizer section of our menu. This riff on that classic dish infuses the spice and smoke of Chipotle Vegenaise while maintaining the natural juicy-meatiness of the mushrooms and veggies by sautéing them directly in the Vegenaise. Much of the heat from the chipotle will dissipate while cooking, but you can adjust your spice level according to how much Fiesta Shreds you include.

SERVES 2

2 tablespoons Chipotle Vegenaise

½ onion, chopped

½ green bell pepper, seeded and thinly sliced

½ red bell pepper, seeded and thinly sliced

½ pound button or cremini mushrooms, cleaned and sliced

1 teaspoon sea salt

¼ teaspoon cayenne pepper

½ teaspoon freshly ground black pepper

1 garlic clove, crushed and minced

2 flour tortillas

¾ cup Follow Your Heart Fiesta Blend Shreds

Fresh cilantro leaves for garnish

Avocado Smash (page 153)

Heat the Chipotle Vegenaise in a large skillet over medium heat, and sauté the onion for 5 minutes.

Increase the heat to medium-high and add the bell peppers. Sauté until slightly soft. Add the mushrooms and cook, stirring frequently, until the mushrooms are tender. Then, add the salt, cayenne, and black pepper and stir to combine. Add the garlic and sauté for 1 minute.

Place one tortilla in a separate skillet over medium heat and add half of the Fiesta Blend Shreds. Heat until slightly melted.

Add the sautéed vegetables to the tortilla and top with the remaining Fiesta Blend. Then, place the second tortilla on top and heat the quesadilla until all of the shreds have melted, carefully flipping once. Serve warm and garnish with cilantro and Avocado Smash.

Sriracha Tofu Buddha Bowl

Marinating tofu is the best way to infuse flavors, as the neutral soy taste can absorb so much. My son David loves sriracha, and if you're like him, be sure to keep the bottle nearby for extra kick. If you can't marinate overnight, cut your tofu into smaller pieces, poke with a fork in a few places, and marinate for at least 30 minutes.

SERVES 2

Sriracha Tofu

¼ cup Sriracha Vegenaise

1 teaspoon toasted sesame oil

1 teaspoon tamari

1½ teaspoons rice vinegar

One 10-ounce container extra-firm tofu, pressed and cut into 8 long, thin pieces

Marinated Kale

1 bunch curly kale, stemmed and chopped (2 cups)

1½ teaspoons toasted sesame oil

1 tablespoon tamari

1 tablespoon rice vinegar

To Assemble

3 cups cooked forbidden rice or brown rice

1 cup shredded carrot

1 cup shredded red cabbage

½ avocado, pitted, peeled, and thinly sliced

Black sesame seeds for garnish

Fresh cilantro for garnish (optional)

Sriracha Vegenaise for topping (optional)

Prepare the Sriracha Tofu: Whisk together the Sriracha Vegenaise, toasted sesame oil, tamari, and rice vinegar in a medium bowl. Add the tofu and marinate overnight.

Preheat the oven to 375°F. Transfer the tofu to a baking sheet and bake for 15 to 20 minutes, until lightly browned. Remove from the oven and let cool.

Prepare the kale: Combine the kale, toasted sesame oil, tamari, and rice vinegar in a large bowl. Toss and massage the kale until all the ingredients are incorporated.

Assemble two bowls by dividing equal amounts of the rice, carrot, cabbage, avocado, kale, and tofu between them. Sprinkle with black sesame seeds and cilantro (if using). Serve and enjoy with an additional dollop of Sriracha Vegenaise, if you like spice.

Jackfruit Carnitas Tacos with Chipotle Lime Crema

We're lucky to have tacos available on almost every corner here in LA, so it's no wonder we have so many in this book. I don't often use jackfruit at home, but recommend it often for those seeking a new meat alternative. Many don't yet know about this versatile ingredient, but our friends at Upton's Naturals have led the way in bringing this key ingredient to the US mainstream. These tacos—crisp and juicy with a little bit of heat—are inspired by a traditional carnitas recipe that a co-worker made in the office a few years ago. You can also save yourself some time by seeking out the Upton's version in stores!

SERVES 4

Spicy Pickled Onions

1 red onion, thinly sliced

1 habanero pepper, seeded and thinly sliced

Juice of 4 limes

Salt

Chipotle Lime Crema

½ cup Chipotle Vegenaise

2 tablespoons fresh lime juice

¼ teaspoon freshly ground black pepper

Salt

Marinated Jackfruit Carnitas

1 teaspoon ground cumin

2 teaspoons dried oregano

1 teaspoon paprika

One 14-ounce can young jackfruit

2 teaspoons olive oil

½ sweet onion, chopped

¼ cup fresh orange juice

1 garlic clove, crushed and minced

To assemble

4 soft corn tortillas

Fresh cilantro for garnish (optional)

Sliced radish for garnish (optional)

4 lime wedges

continued ▶

Prepare the pickled onions: Mix together the red onion, habanero, lime juice, and salt in a small bowl and store in the refrigerator for up to 2 days.

Prepare the jackfruit: Rub the cumin, oregano, and paprika into the jackfruit until evenly coated. Place the olive oil, jackfruit, and chopped onion in a large cast-iron skillet and sauté for 5 to 7 minutes until onions become translucent. Add the orange juice and let reduce until the liquid is almost gone. Add the garlic and cook for 1 minute more. Place the skillet under the broiler and crisp the mixture for 5 minutes. Remove from the broiler and set aside.

Prepare the crema: Whisk together the Chipotle Vegenaise, lime juice, and black pepper in a medium bowl. Add salt to taste and set aside.

Assemble the tacos: Heat the tortillas in a skillet over high heat.

Layer the jackfruit carnitas and pickled onions in the tortillas and garnish with the cilantro and radish (if using). Drizzle the crema over the tacos and serve with the lime wedges. ¡Buen provecho!

Vegan Spicy Tuna Hand Rolls

As a longtime vegetarian since the '70s, I never had a spicy tuna roll. So, I reached out to some of our newly vegan friends to help me with this one. The spice from the sriracha pairs so nicely with the toasted sesame oil and dulse flakes. And I love the kick of the Wasabi Vegenaise.

YIELDS 6 TO 8 CAKES

Spicy Tuna

3 to 4 large Roma tomatoes

1½ teaspoons dulse flakes or aji nori furikake (seasoning mix)

½ teaspoon liquid aminos

1 tablespoon Sriracha Vegenaise

¼ teaspoon minced fresh ginger

1½ teaspoons toasted sesame oil

Wasabi Vegenaise

½ cup Vegenaise

2 teaspoons wasabi powder

Rice

2 cups plus 1 tablespoon water

1 cup uncooked short-grain brown rice, rinsed

1 tablespoon rice vinegar

½ teaspoon sea salt

2 teaspoons evaporated cane sugar

To Assemble

8 nori sheets, cut in half

1 Persian cucumber, julienned

1 carrot, peeled and julienned

2 avocados, pitted, peeled, and sliced

Pickled ginger

Soy sauce

Prepare the Spicy Tuna: Bring a medium pot of water to a boil and prepare a separate bowl of ice water.

Slice a shallow X in the ends of the tomatoes, not the stem side.

Using a slotted metal spoon, add tomatoes to the boiling water and boil for just 25 to 30 seconds. You don't want to cook them, only make them easy to peel. Immediately transfer the tomatoes to the bowl of ice water.

Once the tomatoes are cool, peel off the skin and cut the tomatoes into quarters. Scoop out and discard the seeds and aggressively squeeze out any excess water. The tomatoes should not be overly wet.

Place the tomatoes in a medium bowl and smash with a fork or potato masher until mostly smashed. Strain the tomatoes of excess liquid. Mix with the dulse flakes, liquid aminos, Sriracha Vegenaise, ginger, and toasted sesame oil. Refrigerate until ready to use, up to 1 day. Strain out any excess liquid before building the hand rolls.

continued ▶

Prepare the Wasabi Vegenaise: Whisk together the Vegenaise and the wasabi powder in a small bowl.

Prepare the rice: Bring water and rice to a boil in a medium saucepan over high heat. Reduce heat and simmer for 45 minutes, not disturbing the lid until the entire cook time is complete. Remaining undisturbed, remove from heat, and let sit for 10 minutes. Fluff with fork and add the rice vinegar, salt, and sugar and mix in. Mash the rice with a fork until a sticky texture is achieved. If using white rice, don't mash.

Assemble the hand rolls: Using dry hands, position one halved nori sheet horizontally in front of you, shiny side down. Spread a thin layer of rice on the left half of the nori sheet and top with a scoop of "tuna." Drizzle some Wasabi Vegenaise over the "tuna." Position your veggies in a diagonal line angled from the top left down to the bottom right of the rice. Starting with the bottom left corner of nori, roll from the left to right into a cone, using the bottom center of the nori sheet as the bottom point of your cone. Seal the roll closed with a bit of water or piece of rice.

Serve with the pickled ginger and soy sauce.

Sneaky Healthy Shepherd's Pie

When our kids were young, a favorite recipe at home was our Kasha Potato Pie, which we even occasionally served as a restaurant special. It was a baked, layered casserole with a toasted kasha base, topped with sautéed onions, mushrooms, and mashed potatoes. This recipe is a bit more vegetable heavy and a perfect way to sneak veggies and nutrition into a meal that is bound to become a family favorite in your home as well.

SERVES 4 TO 6

Mash Topping

6 cups quartered red potatoes (3 pounds), peeled if desired

1 small head cauliflower, stemmed, separated into florets

½ cup Vegenaise

3 tablespoons vegan butter

Sea salt and freshly ground black pepper

Filling

1 tablespoon neutral-tasting oil (safflower or grapeseed)

1 small white onion, chopped

½ cup diced carrot (2 medium carrots)

1 small head broccoli, stemmed and roughly chopped

One 12-ounce package ground veggie crumble

1 cup fresh or frozen corn kernels

1 cup fresh or frozen green peas

2 garlic cloves, crushed and minced

1 tablespoon tomato paste

1 teaspoon dried thyme

¼ cup veggie broth

1 teaspoon vegan Worcestershire sauce

1 teaspoon water

Sea salt and freshly ground black pepper

Prepare the mash topping: Boil 6 cups of water in a large pot. Add the potatoes and cauliflower and cook until fork-tender.

Transfer to a large bowl and mash with the Vegenaise, vegan butter, and salt and black pepper to taste.

Prepare the filling: Preheat the oven to 400°F.

Heat the oil in a large skillet over medium heat and sauté onion, carrot, and broccoli for 5 to 7 minutes, until the onion is translucent and the carrot is barely tender. Add the ground veggie crumble, corn, peas, garlic, tomato paste, thyme, veggie broth, and Worcestershire sauce and sauté until the liquid is reduced, 10 to 15 minutes. Deglaze the pan with 1 teaspoon of water. Season with salt and black pepper to taste. Remove from the heat and set aside.

Fill a casserole dish or large cast-iron skillet with the filling and smooth with a spatula. Layer with the mash topping and smooth. Place the casserole on a baking sheet and bake for 20 to 25 minutes, until golden brown. Remove from the oven and let cool for at least 15 minutes before serving.

DIPS & SAUCES

Basic Garlic Aïoli

A secret in our house is that we almost exclusively use our Organic Garlic Aïoli Vegenaise, not Original Vegenaise, as one might think. After decades of sticking with the Original, I finally had to admit to myself that I liked the Garlic Aïoli even more. This is a basic make-at-home version if you can't access our Garlic Aïoli Vegenaise on shelf. Three ingredients. Endless recipes with which to pair.

YIELDS 1 CUP

1 cup Vegenaise

1 tablespoon crushed and minced garlic (from 2 cloves)

1 teaspoon fresh lemon juice

Whisk together all the ingredients in a small bowl.

Horseradish Sauce

Chicago, where I grew up, is a very meaty town, and a condiment commonly used there is horseradish sauce. Although I haven't eaten meat since I left there 50 years ago, I still enjoy a good horseradish sauce, and this one packs quite a punch, especially when added to roasted broccoli or cauliflower steaks.

YIELDS 1¼ CUPS

5 tablespoons prepared horseradish

1 cup Vegenaise

½ teaspoon freshly ground black pepper

1 tablespoon finely chopped fresh chives

Squeeze out as much liquid as you can from the horseradish. Whisk together all the ingredients in a small bowl.

From front to back: Horseradish Sauce,
Curry Mayo, and Béarnaise Sauce

Curry Mayo

When I first tasted this recipe, I became obsessed with it, and I immediately knew that my youngest son, Aaron, with his love of Indian food, would go crazy for it, too. By simply combining Vegenaise, curry powder, and lime juice, we unlocked an explosion of flavor. Then, we added a kick of cayenne and this sauce became unbeatable. Try it and see if you don't agree.

YIELDS 1 CUP

1 cup Vegenaise

2 teaspoons curry powder

1 tablespoon fresh lime juice

⅛ teaspoon cayenne powder

Paprika for garnish

Mix together all the ingredients, except the paprika, in a medium bowl, then top with the paprika.

Béarnaise Sauce

Considered a daughter sauce to the traditional French hollandaise "mother sauce," our vegan version of Béarnaise elegantly and classically showcases the shallot and the herb tarragon, which really shine with Vegenaise as a base. This recipe is a great illustration of the versatility that Vegenaise provides in sauces of every kind and relieves the stress of trying to create an emulsion at home.

YIELDS 1½ CUPS

3 tablespoons finely minced shallot

1 tablespoon vegan butter

1 tablespoon white wine vinegar

1 cup Vegenaise

2 to 3 teaspoons fresh lemon juice

1 tablespoon finely chopped fresh tarragon

½ teaspoon freshly ground black pepper

⅛ teaspoon ground turmeric

Sauté the shallot in the vegan butter in a small skillet over medium heat for 1 minute. Lower the heat to low, add the white wine vinegar, and simmer until the vinegar has evaporated. Cook the shallot for an additional 5 minutes, or until translucent. Remove from the heat and let cool.

Place the Vegenaise in a small bowl. Add the cooled shallot and fold in the remaining ingredients with a fork, starting with 2 teaspoons of the lemon juice. Add more lemon juice until the sauce is creamy and not quite pourable, but be careful not to overmix.

Paprika Mayo

YIELDS 1 CUP

Paprika is commonly used in Hungarian and Central American cuisine and is a spice that we most often use in baked dishes and stews. I find that the regional differences are so subtle, yet significant, that many local spice stores have entire shelves of different paprika vials. It is a generous component of the Follow Your Heart kitchen's chili, once a rotating soup special, but now, by popular demand, served daily in the restaurant. This sauce is an excellent coating for roasting chickpeas or as a nice complement to roasted corn or winter veggies.

1 cup Vegenaise

1½ teaspoons red wine vinegar

2 garlic cloves, crushed and minced

¼ cup canned fire-roasted tomatoes

1 tablespoon paprika

Sea salt and freshly ground black pepper

Chopped fresh parsley (optional)

Combine the Vegenaise, vinegar, garlic, fire-roasted tomato, and paprika in a food processor or blender and blend until smooth. Season with salt and black pepper to taste. Transfer to a serving bowl and top with parsley (if using).

Quick Tartar Sauce

YIELDS 1½ CUPS

As long as they're not from real crab, I'm a huge fan of "crab" cakes and order them whenever I find them on a menu. Most often, they are served with tartar sauce. This is a simple and quick version of the Tartar Sauce we pair with our famous Follow Your Heart restaurant's corn fritters and also makes a great shortcut for use in an easy chickpea "tuna" salad.

1 cup Vegenaise

½ cup dill pickle relish, drained of excess liquid

½ teaspoon freshly ground black pepper

¼ teaspoon garlic powder

¼ teaspoon vegan Worcestershire sauce

½ teaspoon fresh lemon juice

Whisk together all the ingredients in a small bowl. Refrigerate for 30 minutes prior to using.

From front to back: Avocado
Smash, Paprika Mayo, and
Quick Tartar Sauce

Avocado Smash

We've had many ardent fans over the years, so when we first heard of a little handmade zine called *Vegenaise with a Vengeance*, I knew I needed copies for the whole office. When they arrived with their black-and-white print with stapled pages, we were beyond flattered. Inspired by the author, Rudy Ramos, a.k.a. Vegicano, and his beautiful zine, this spin on guacamole is addictive and not to be confused with our Avocado Oil Vegenaise. You'll enjoy the increased kick of cayenne and be tempted to use it on just about everything you can.

YIELDS 2 CUPS

4 large avocados, pitted and peeled

1 cup Vegenaise

2 tablespoons fresh lime juice

2 tablespoons cider vinegar

⅛ teaspoon cayenne pepper

Sea salt and freshly ground black pepper

Place the avocado flesh in a small bowl and mash with a fork until smooth.

Whisk in the Vegenaise, lime juice, vinegar, and cayenne. Add salt and black pepper to taste.

Front: Rémoulade;
Back: Hollandaise Sauce

Hollandaise Sauce

The Follow Your Heart sandwich counter was originally open from 11 to 3, and only served lunch. When we first launched our breakfast menu in the late '80s, Tofu Benedict was an early favorite. We created our own vegan version of the classic French "mother sauce" that has topped those Benedicts to this day. While not our exact hollandaise recipe, this is a quick and easy option that you can make at home, and is also an amazing accoutrement to grilled asparagus or broccoli.

YIELDS 1¼ CUPS

1 cup Vegenaise

¼ cup vegan butter, melted

1½ tablespoons fresh lemon juice

⅛ teaspoon ground turmeric

Pinch of cayenne pepper

Whisk together all the ingredients in a small bowl.

Rémoulade

Our rémoulade is tangy, spicy, smoky, herbaceous, and all-around amazing. I think of it as an amped-up, smoky version of our retail Thousand Island, one of our classic salad dressing offerings that is now sold internationally. You might find yourself licking the spoon on this one.

YIELDS 1½ CUPS

1 cup Vegenaise

1 teaspoon Dijon mustard

¼ cup finely chopped cornichons or dill pickle relish, finely chopped

1 large garlic clove, crushed and minced

1 tablespoon smoked paprika

2 teaspoons finely chopped fresh tarragon

1 teaspoon capers, drained and chopped

1 teaspoon vegan Worcestershire sauce

¼ teaspoon cayenne pepper

Whisk together all the ingredients in a small bowl. Refrigerate for at least 30 minutes prior to using.

Cajun Sauce

Inspired by a trip to New Orleans a few years ago to attend a tradeshow for food scientists, our R&D team challenged themselves to identify every spice and flavor in some of the local dishes. Although we may not have nailed every spice, we did end up creating this sauce, which we mostly use on our Crispy Mushroom Po' Boy Sandwiches (page 77).

YIELDS 1¼ CUPS

1 cup Vegenaise

2 tablespoons ketchup

1 tablespoon prepared horseradish

2 teaspoons onion powder

2 teaspoons paprika

1 teaspoon crushed and minced garlic

½ teaspoon freshly ground black pepper

½ teaspoon dried oregano

½ teaspoon dried thyme

¼ teaspoon cayenne pepper

Whisk together all the ingredients in a small bowl.

Sweet Fry Sauce

An added benefit to having so many foodie people around me is the recipe inspiration and cultural education that I get on an almost daily basis. One local example is fry sauce, which was introduced to me a few years ago and which I fully agree is a great excuse for making and dipping French fries. Originating in Utah, there are many takes on fry sauce, and this one brings a sweetness and kick that I find irresistible.

YIELDS 1½ CUPS

1 cup Vegenaise

½ cup ketchup

2 teaspoons vegan Worcestershire sauce

1 tablespoon evaporated cane sugar

1 teaspoon freshly ground black pepper

Whisk together all the ingredients in a small bowl.

Vegenaise de Provence

Years ago, on our honeymoon, my wife, Kathy, and I traveled through the South of France, where the abundance of fresh herbs in home gardens and restaurant patios inspired countless recipes. Since then, we've maintained this practice at home, whether in the garden or just in pots. Using fresh herbs whenever the opportunity arises is a pleasant reminder of that memorable and magical time, enjoying each other's company, the unforgettable sights, and some amazing food.

YIELDS 1¼ CUPS

2 fresh mint leaves

Leaves from 3 sprigs thyme

Leaves from 2 sprigs tarragon

Leaves from 2 sprigs marjoram

Leaves from 2 sprigs oregano

Leaves from 1 sprig rosemary

1 cup Vegenaise

1 teaspoon minced fresh chives

½ teaspoon fresh lemon juice

Finely chop the mint, thyme, tarragon, marjoram, oregano, and rosemary. Combine the Vegenaise, herbs, chives, and lemon juice in a small bowl and whisk together. Refrigerate for 1 hour prior to using.

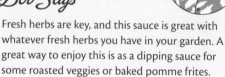

 Bob Says

Fresh herbs are key, and this sauce is great with whatever fresh herbs you have in your garden. A great way to enjoy this is as a dipping sauce for some roasted veggies or baked pomme frites.

Creamy Chimichurri Sauce

YIELDS 2½ CUPS

In the late '80s, we needed a name for the new company we created to pro-duce the deli salads and dressings that we sold. Inspiration came from the con-cept articulated by R. Buckminster Fuller that we are all travelers on "Spaceship Earth." I wanted to connect that to what we were doing, and we expected salad dressings to be a big part of our business. With Thousand Island being one of the more popular dressings of the time, "Earth Island" came to mind as a good choice for a name. Our thought was that the dressings we made would draw upon recipes from all around the world, and although that initial concept has shifted somewhat over the years, we're still inspired by world cuisine and by the vision that Bucky Fuller had of Earth as the singular and precious home that we all share. Chimichurri sauce, a favorite around our office, is standard Argentinean fare and this one packs a punch that is perfect on seitan and cauliflower steaks.

½ sweet onion, chopped

5 garlic cloves, roughly chopped

¼ cup cider vinegar

1½ tablespoons red pepper flakes

½ cup packed fresh cilantro leaves

½ cup packed fresh parsley leaves

1 teaspoon sea salt

⅔ cup olive oil

1 cup Vegenaise

Pulse together all the ingredients, except the Vegenaise, in a food processor. Transfer to a medium bowl and refrigerate for 15 minutes. Gently fold in the Vegenaise with a fork. Refriger-ate for 30 minutes if you prefer a thicker sauce.

Ultimate Artichoke Dipping Sauce

We have an organic garden on our production facility grounds, and this summer, we harvested eight beautiful, little artichokes. One of our long-standing employees, Katie, cooked them up in our tiny office kitchen and whipped together this sauce to dunk them in. It truly is the ultimate artichoke dipping sauce.

YIELDS 1¼ CUPS

1 cup Vegenaise

1 teaspoon lemon zest

2 tablespoons fresh lemon juice

1 garlic clove, crushed and minced

⅛ teaspoon ground turmeric

2 teaspoons freshly ground black pepper

1 teaspoon sea salt

Whisk together all the ingredients in a small bowl. Refrigerate for 30 minutes prior to using.

Front: Garlic Parmesan Sauce;
Back: Marsala Wine Sauce

Garlic Parmesan Sauce

Destined for garlic bread, this sauce is also an exceptional dip for vegetables or spread on some gourmet locally baked bread. Our Parmesan has become an instant best seller, and really shines in this sauce.

YIELDS 1¼ CUPS

1 cup Vegenaise

¼ cup Follow Your Heart Shredded Parmesan

3 garlic cloves, crushed and minced

1 tablespoon fresh lemon juice

1 teaspoon finely chopped fresh parsley

Whisk together all the ingredients in a small bowl.

Marsala Wine Sauce

In the early 2000s, shortly after the launch of Grapeseed Oil Vegenaise, I brought some home and created this elegant sauce. It quickly became a regular drizzle for steamed or roasted vegetables, such as broccoli, cauliflower, or asparagus. I find that the flavor of Grapeseed Oil Vegenaise yields a subtle nutty hint that is missed when I use Original or our other Vegenaise varieties.

YIELDS 1 CUP

1 cup Grapeseed Oil Vegenaise

2 tablespoons dry marsala wine

2 tablespoons water

Combine the Vegenaise and marsala wine in a small pot or saucepan over medium-low heat, stirring constantly. Then, add the water, 1 teaspoon at a time, until the sauce dribbles off a spatula but has not separated.

Bob Says

The trick to this sauce is the slow and low heat. You don't want to break the emulsion; just slowly stir and add water until you have a nice creamy yet pourable sauce.

Your Very Own Secret Sauce

It's time for you to be an adventurer in the kitchen. With these basic ingredients as a jumping-off point, and at quantities that you decide upon, let your inspiration fly and prepare to be locally famous with your very own Secret Sauce.

YIELDS 1½ TO 2 CUPS

1 cup Vegenaise

_____ pickle relish (sweet or dill or bread & butter)

_____ ketchup or tomato paste

_____ paprika (smoked or regular)

_____ garlic (powder or fresh)

_____ onion (powder or fresh)

_____ spice (cayenne or hot sauce)

_____ herbs of your choice (dried or fresh)

_____ other secret ingredients

Whisk together the Vegenaise and your choice of additional ingredients in a small bowl.

Bob Says

The trick to a good secret sauce is not adding too much liquid, keeping it thick. Strain any ingredients of their liquid and gently fold in, taking care not to overmix.

Front: Harissa Aïoli;
Back: Herbed French Onion Dip

Harissa Aïoli

This Harissa Aïoli is another recipe inspired by the idea that Earth Island products would incorporate regions of the world not always well represented. While we've incorporated harissa into select dinner specials over the years, this rich, smoky, and deeply spiced sauce is based off the trio of pan-toasted seeds, and pairs so nicely with eggplant or hummus.

YIELDS 1¼ CUPS

Harissa Paste

1 cup boiling water

2 ounces dried chiles de arbol

1 ounce dried ancho chiles

2 teaspoons coriander seeds

2 teaspoons cumin seeds

1 teaspoon caraway seeds

3 sun-dried tomatoes, chopped

1 teaspoon smoked paprika

3 large garlic cloves, crushed

1 teaspoon fresh lemon juice

1 tablespoon tomato paste

Kosher salt

2 to 3 teaspoons extra-virgin olive oil

Aïoli

1 cup Vegenaise

2 garlic cloves, crushed and minced

Prepare the harissa paste: Place the boiling water and chiles in a metal bowl and cover. Soak for 20 minutes.

Toast the coriander, cumin, and caraway seeds in a dry skillet, moving them often so as not to burn. Remove from the heat and set aside to cool. Grind in a mortar and pestle or spice grinder.

Drain the chiles. Cut off the stems and remove and discard the seeds.

Combine the chiles, tomatoes, toasted seeds, paprika, garlic, lemon juice, tomato paste, and salt to taste in a food processor. With the processor on low speed, slowly add the oil until a thick paste is formed. Transfer the harissa paste to a small jar and store, refrigerated for 3 days or freeze for later

Prepare the aïoli: Mix ¼ cup of the harissa paste with the Vegenaise. Mix the minced garlic with the sauce.

Bob Says

Harissa paste is also available retail if you want to skip the step of creating your own harissa. I find it in the international aisle at my local grocery store.

Herbed French Onion Dip

When I first tasted this herbed onion dip, it instantly recalled family holiday memories when we always had a bowl of onion dip on the living room table for snacking. I consider this a grown-up version of that classic onion dip, which was admittedly mixed from a packet. I recommend opting for fresh vegetable crudités if you want to stay in the healthier realm.

YIELDS 1½ CUPS

3 tablespoons olive oil

1 large onion, chopped

2 sprigs thyme

2 sprigs tarragon

1 sprig rosemary

1 cup Vegenaise

2 tablespoons finely chopped fresh parsley

1 garlic clove, crushed and minced

1 shallot, finely chopped

1 teaspoon freshly ground black pepper

Heat the oil in a large skillet or heavy-bottomed pot over medium heat. Add the onion and cook for 5 minutes, stirring frequently to avoid burning. Lower the heat and add the herbs. Place the lid on the skillet slightly askew for steam to escape. Cook over medium-low heat for 30 to 35 minutes, until the onion is softened and starting to caramelize. Remove from the heat and remove and reserve the herbs. Set aside and let cool. When cool, remove the leaves from the herbs and use for garnish or save for a freezer broth.

Combine the Vegenaise and parsley in a medium bowl and fold in the onion mixture. Add the garlic, shallot, and black pepper and mix well.

DESSERTS

Dreamy Carrot Cake with Vegan Cream Cheese Frosting

We sell a lot of carrot cake at Follow Your Heart in both the restaurant and the store and, honestly, we have a hard time keeping it on shelf. Over the years, we've learned that Vegenaise is a great baking ingredient. Somehow, its components all combine to maintain moisture and lift while ensuring that the distinct flavors of those core ingredients all come through.

YIELDS ONE 2-LAYER
8-INCH ROUND CAKE;
SERVES 6 TO 8

Carrot Cake

Nonstick spray or cooking oil for pans

2½ cups unbleached all-purpose flour

1½ teaspoons baking powder

1 teaspoon baking soda

1½ teaspoons ground cinnamon

1 teaspoon ground ginger

½ teaspoon ground nutmeg

⅛ teaspoon ground cloves

½ teaspoon sea salt

1 cup Follow Your Heart Plain Dairy-Free Yogurt

¼ cup cornstarch

½ cup packed light brown sugar

1½ cups evaporated cane sugar

2 teaspoons vanilla extract

1 cup Vegenaise

3 cups peeled and shredded carrot, (1 pound, or 6 to 7 medium carrots)

1½ cups nuts (optional)

1 cup raisins (optional)

Frosting

¼ cup vegan butter, at room temperature

¼ cup nonhydrogenated shortening, at room temperature

½ cup Follow Your Heart Dairy-Free Cream Cheese, at room temperature

4½ cups powdered sugar

2 teaspoons vanilla extract

For decorating (optional)

Natural orange food coloring

Marzipan

Fresh mint leaves

continued ▶

Prepare the cake: Preheat the oven to 350°F. Spray two 8-inch round cake pans with nonstick spray or cooking oil and line the bottoms with parchment paper, then spray again.

Whisk together the flour, baking powder, baking soda, cinnamon, ginger, nutmeg, cloves, and salt in a medium bowl and set aside.

Combine the yogurt and cornstarch in a large bowl and mix with a fork or small whisk until well blended (a few clumps are fine). Add the brown and cane sugar and beat with an electric mixer on medium-high speed until well combined. Add the vanilla and blend for a few more seconds. Add the Vegenaise and beat on the lowest speed until just combined.

Add half of the flour mixture and mix with spatula or spoon to incorporate, then repeat with the remaining flour mixture. Add the shredded carrot and mix until well distributed throughout the batter. Add the nuts and raisins (if using) and mix well.

Divide evenly between the prepared pans and bake for 35 minutes, or until a toothpick inserted into the center of a cake layer comes out clean. (If using nuts or raisins, add an additional 5 minutes to the baking time, testing at 35 minutes.) Remove from the oven and let the layers cool for at least 30 minutes in their pan before inverting onto a wire rack and letting cool completely (at least an hour). If not frosting the cakes immediately, wrap in plastic wrap and store refrigerated until ready to use.

Prepare the frosting: Combine the vegan butter, shortening, and cream cheese in a large bowl and cream together with an electric mixer on medium-high speed for about 1 minute until well-combined. Add the powdered sugar in 1-cup increments and mix after each addition at high speed until smooth and creamy (at least 5 minutes), then add the vanilla and beat in. Store tightly covered and refrigerated until ready to use.

To frost the cake, place one layer on a flat serving surface and cover with a generous layer of icing. Stack the second layer on top and cover the top and sides with the remaining icing.

If desired, to make marzipan "carrots," work orange food coloring into marzipan and shape into carrot shapes, then gently press down on them several times with a sharp knife to create lines. Arrange eight "carrots" on the cake and add small fresh mint leaves to create their leaves.

Lemon Poppy Seed Cake with Lemon Glaze

So many lemons fall from the fruit trees on the hill above our house that we sometimes make up excuses to use them. Whether you're in the same predicament or bought a bag from your local farmers' market, this light cake is a great way to use some. Without the heaviness of a butter-based cake, this recipe moonlights as a brunch indulgence that you can serve as a morning "dessert."

YIELDS ONE 9-INCH ROUND OR BUNDT CAKE; SERVES 8 TO 10

Cake

Nonstick spray for pan

1 cup evaporated cane sugar

Zest of 2 lemons

½ cup Vegenaise

½ teaspoon vanilla extract

¾ cup nondairy milk

½ teaspoon white vinegar

3 to 4 tablespoons lemon juice (from 2 lemons)

1¾ cups unbleached all-purpose flour

1½ teaspoons baking powder

¼ teaspoon baking soda

½ teaspoon sea salt

1 tablespoon poppy seeds

Lemon Glaze

3 to 4 tablespoons fresh lemon juice (from 2 lemons)

2 cups powdered sugar

Prepare the cake: Preheat the oven to 350°F. Spray a 9-inch cake pan or Bundt pan with nonstick spray and set aside.

Combine the cane sugar and lemon zest in a large bowl and stir together to create a mixture that resembles damp sand. Stir in the Vegenaise and vanilla until smooth.

Stir together the milk and vinegar in a liquid measuring cup and let sit for 2 minutes. Add the lemon juice and set aside.

Whisk together the flour, baking powder, baking soda, and salt in a medium bowl to combine.

Add the vinegar mixture to the sugar mixture and stir to combine. Add half of the flour mixture and mix with spatula or spoon to incorporate, then repeat with remaining flour mixture. Do not overmix. Gently stir in the poppy seeds to evenly incorporate.

Pour the batter into the prepared pan and bake 40 minutes, or until a toothpick inserted into the center of the cake comes out clean. Remove from the oven, let cool for 15 minutes in the pan, then transfer to a wire rack to cool completely.

continued ▶

Prepare the lemon glaze: Whisk together 3 tablespoons of the lemon juice and the powdered sugar in a small bowl until smooth, gradually adding more lemon juice as needed until the glaze is thick but still pourable.

When the cake is completely cool, set on a wire rack over a baking sheet and drizzle the glaze over the cake. Let cool for 1 hour prior to serving.

Blender Blackout Double Chocolate Cake

I'm always flattered when other chefs find unique and new uses for Vegenaise and are inspired in their own kitchens. Chef Olivia, who was first introduced to us by Danny Seo, wowed us with this seemingly simple yet incredibly rich cake recipe and I knew I needed to include it in this collection. In this recipe, Vegenaise plays many roles. It helps coat the pan, provides moisture and lift to the batter, and keeps the sauce smooth for when you spread it over the baked and cooled cake.

YIELDS ONE 9-INCH
ROUND CAKE

Vegenaise and unsweetened cocoa powder for pan

2 cups self-rising flour

1 cup coconut sugar

½ cup unsweetened cocoa powder

1¼ cups Original Vegenaise

2 teaspoons vanilla extract

1 cup oat milk

Two 4-ounce vegan dark chocolate bars

Preheat the oven to 350°F.

Line the bottom of 9-inch cake pan with parchment paper. Brush the inside of the pan with Vegenaise and dust with cocoa powder.

Combine the flour, coconut sugar, and cocoa powder in a blender and process until well mixed. Add 1 cup of the Vegenaise plus the vanilla and oat milk and blend just until a batter forms, scraping down the sides if necessary.

Pour the batter into the prepared pan and bake for 30 minutes, or until a toothpick inserted into center of the cake comes out clean. Remove from the oven, let cool completely, then remove the cake from the pan.

Use a peeler to make shaved ribbons from ½ of one of the chocolate bars and set aside.

Break up the remaining 1½ bars of chocolate and melt over a double boiler. Remove from the heat, let cool for 30 seconds, then stir in the remaining ¼ cup of Vegenaise.

Spread the melted chocolate mixture over the cooled chocolate cake. Let set for 5 minutes in the fridge. Garnish with the chocolate shavings.

Banana Bread and Toasted Coconut Waffles

Lazy Sundays are good for many things and a great occasion for one of my all-time favorites, waffles. My first experience with banana coconut waffles was in a restaurant in Kauai and I haven't been able to get them out of my head. I think you'll find this recipe to be equally memorable and one that far exceeds an ordinary, prenoon level of indulgence.

YIELDS 3 CUPS BATTER, OR 6 WAFFLES

1½ cups self-rising flour

2 teaspoons ground cinnamon, plus more for serving

¼ cup Vegenaise, plus more for waffle iron

4 ripe bananas

¼ cup pure maple syrup, plus more for serving

½ cup chopped walnuts, plus more for serving

⅓ cup filtered water

¼ cup unsweetened coconut flakes

Combine the flour, cinnamon, Vegenaise, two of the bananas, the maple syrup, ¼ cup of the walnuts, and the filtered water in a blender and blend until smooth. Add the remaining walnuts to the batter and fold in.

Preheat a waffle iron according to the manufacturer's instructions and brush inside lightly with Vegenaise.

Pour in the batter and cook for 2 to 4 minutes, until the waffles are golden and cooked through. Repeat with the remaining batter.

Place the coconut flakes in a dry skillet and toast over high heat, stirring constantly for approximately 1 minute, or until golden.

Slice the remaining two bananas into spears and roll in the toasted coconut flakes.

Plate the waffles and top with the sliced banana, cinnamon, the remaining walnuts, and maple syrup.

Vegenaise Sugar Cookies

Cookies? Vegenaise? You bet! We've heard it said over and over again that "Vegenaise just makes it better," and this recipe is no exception. When there's a plate of these cookies around—don't take too long thinking about when to dive in—they might be gone.

YIELDS 32 COOKIES

Neutral-tasting oil (safflower or grapeseed) for greasing

1 cup evaporated cane sugar

2 cups unbleached all-purpose flour

1 teaspoon baking soda

Pinch of sea salt

1 cup Vegenaise

1 teaspoon vanilla, lemon, or almond extract

2 tablespoons coarse sugar or vegan sprinkles (optional)

Preheat the oven to 350°F.

Stir together the cane sugar, flour, baking soda, and salt in a medium bowl.

Mix together the Vegenaise and vanilla in a large bowl. Add the flour mixture and mix together until fully incorporated. The dough will be crumbly.

Shape into walnut-size balls. Place 2 inches apart on a large parchment-lined baking sheet and mash with a fork to ½-inch thickness. Sprinkle with coarse sugar or sprinkles (if using).

Bake for 10 to 12 minutes, until they look dry and lightly golden. Remove from the oven and transfer to a wire rack to cool.

Strawberry Scone Cake with Lemon Whip

Every year, on Mother's Day, we host a special brunch where every mom gets a long-stemmed rose and they're treated to an amazing meal that's meant to reward and recognize them for their love and hard work. This recipe is perfect for such an occasion. Is it a scone or a cake or something extra? With this bright, springtime recipe, it's all of the above and together they synthesize into an amazing dessert, teatime indulgence, or morning treat.

YIELDS ONE 2-LAYER 8-BY-6-INCH CAKE

2 cups self-rising flour

1 cup coconut sugar, plus 2 tablespoons for whip (optional)

½ cup Vegenaise

1 cup freeze-dried strawberries, plus more for garnish

Zest and juice of 1 lemon

2 cups chilled coconut cream

2 cups fresh strawberries

Preheat the oven to 350°F. Line an 8-by-12-inch baking dish with parchment paper.

Stir the flour and coconut sugar together in a large bowl. Use a fork to mash the Vegenaise into the flour mixture until its consistency resembles that of wet sand.

Use your hands to break up the freeze-dried strawberries into the flour mixture.

Set aside the zest for the coconut whip. Add the lemon juice to the flour mixture and use a spatula to combine.

Spread the batter into the prepared pan. Bake for 25 minutes, or until golden brown and cooked through.

Remove from the oven and let cool in the refrigerator for 30 minutes.

Prepare the lemon whip: Combine the lemon zest and chilled coconut cream in a blender or food processor. Process until whipped, scraping down the sides as needed. Add 2 tablespoons of coconut sugar for a sweetened whip, if desired. Cover and refrigerate until ready to use.

Hull and slice the strawberries.

Cut the cake in half crosswise so you have two layers. Before serving, spread the lemon whip over both sections of the cake and top with the sliced fresh strawberries (Note: If done too far in advance, the cake will become soggy.).

Stack the cake layers. Pass the remaining freeze-dried strawberries through a strainer to top the cake with strawberry dust.

Chocolate Coconut Cake

I guarantee this cake will become a favorite request for birthdays. Paul, for many years, would have what we knew as the "King of Coconut Fudge Cake" to celebrate his birthday and this recipe is our representation of that level of decadence in a single bite. Be sure to share this one! The coconut lightens the richness of the chocolate but doesn't make this dessert any less fulfilling.

YIELDS ONE 2-LAYER
8-INCH ROUND CAKE;
SERVES 4 TO 6

Cake

Vegenaise or nonstick spray for pan

2 cups unbleached all-purpose flour

1½ cups evaporated cane sugar

½ cup unsweetened cocoa powder

1 teaspoon baking powder

1 teaspoon baking soda

½ teaspoon sea salt

1½ cups Vegenaise

One 13-ounce can coconut milk

1½ teaspoons vanilla extract

Frosting

1 cup vegetable shortening

½ cup unsweetened cocoa powder

1 teaspoon vanilla extract

5 cups powdered sugar

3 to 4 tablespoons unsweetened nondairy milk

½ cup unsweetened coconut flakes

Prepare the cake: Preheat the oven to 350°F. Prepare two 8-inch round pans by lining the bottoms with parchment paper and coating with a thin layer of Vegenaise or nonstick spray.

Sift the flour, cane sugar, cocoa powder, baking powder, baking soda, and salt into large bowl.

Whisk together the Vegenaise, coconut milk, and vanilla in a separate large bowl.

Gradually mix the flour mixture into the Vegenaise mixture until smooth.

Divide evenly between the prepared pans and bake for 20 to 25 minutes. Remove from the oven, let cool for 20 minutes in the pans, and carefully transfer the layers to a wire rack to cool completely.

Prepare the frosting: Whip together the shortening and cocoa powder in a medium bowl until smooth. Stir in the vanilla and powdered sugar. Slowly pour in the milk until the frosting reaches your desired consistency.

Once fully cooled, frost one cake layer and layer with half of the coconut flakes. Place the other cake layer on top and continue to frost the outside. Top with the remaining coconut flakes.

Chocolate Chip Banana Bread with Cinnamon Streusel Crumble

The first thing we do on our Hawaiian vacations is go to the store and stock up on groceries, including enough of those incredible local bananas to last us until the next farmers' market. They have so many different and distinct varieties that we invariably end up taking home more than we're likely to eat while there. So, our tradition has become that Claire bakes up a loaf of banana bread so we don't waste a single one. Sometimes, we swap out the chocolate chips for toasted pecans, though this more decadent recipe is always first out of the oven. (Note: During the many trips to Mexico in my younger years, I was known by all my friends as Banana Bob.)

SERVES 4 TO 6

Banana Bread

Nonstick spray for pan

2 cups unbleached all-purpose flour

1 cup dark brown sugar

1 teaspoon baking powder

½ teaspoon baking soda

½ teaspoon sea salt

½ teaspoon ground cinnamon

1 teaspoon ground allspice

½ teaspoon ground ginger

1 cup mashed bananas (about 3 bananas)

1 cup unsweetened coconut creamer

½ cup Vegenaise

2 teaspoons cider vinegar

1 tablespoon vanilla extract

1½ cups nondairy semisweet chocolate chips

Cinnamon Streusel Crumble

½ cup packed dark brown sugar

1 teaspoon ground cinnamon

1 tablespoon vegan butter, melted

Preheat the oven to 350°F. Lightly coat a 9-by-5-inch glass loaf pan with nonstick spray.

Prepare the bread: Whisk together the flour, brown sugar, baking powder, baking soda, salt, and spices in a large bowl.

Whisk together the bananas, coconut creamer, Vegenaise, vinegar, and vanilla in a separate large bowl. Add half of the flour mixture and mix with a spatula or spoon to incorporate, then repeat with the remaining flour mixture. Do not overmix. Fold in the chocolate chips.

Prepare the crumble: Mix together the brown sugar and cinnamon in a medium bowl. Stir in the melted butter.

Spread the batter evenly into the prepared loaf pan and sprinkle with the crumble mixture. Bake for 50 to 60 minutes, rotating the pan halfway through baking, until a toothpick inserted into the center of the cake comes out clean. Remove from the oven and let cool for 15 minutes in the pan before transferring to a wire rack to cool completely.

Chocolate Zucchini Loaf

One holiday tradition at Follow Your Heart for many years has been to sell our "Heat & Eat" Thanksgiving and Christmas meals that you can reserve and pick up fresh the day or two before each holiday. Whether you get the whole meal or items à la carte, they make holiday meals easy. Part of the offerings are our famous zucchini and cranberry breads, which sell out each year. Although the ones we make for the Heat & Eat menu are more traditional and savory, this indulgent Chocolate Zucchini Loaf is a perfect end to any holiday meal.

SERVES 4 TO 6

Nonstick spray for pan

½ cup granulated sugar

½ cup light brown sugar

½ cup Vegenaise

One 5.3-ounce container Follow Your Heart Plain Dairy-Free Yogurt

2 teaspoons vanilla extract

½ cup unsweetened nondairy milk

2½ cups unbleached all-purpose flour

¼ cup unsweetened cocoa powder

1 teaspoon baking powder

1 teaspoon baking soda

2 teaspoons ground cinnamon

1 medium zucchini, shredded or grated

¼ cup unsweetened coconut flakes

2 tablespoons sliced almonds

2 tablespoons agave nectar

Preheat the oven to 375°F. Lightly spray a 9-by-5-inch loaf pan with nonstick spray.

Whisk together the granulated and brown sugar and Vegenaise in a large bowl until well combined. Add the yogurt and whisk until the mixture is a uniform slurry. Add the vanilla and nondairy milk and mix until well combined.

Whisk together the flour, cocoa powder, baking powder, baking soda, and cinnamon in a medium bowl until well combined.

Add the flour mixture to the yogurt mixture and mix with a spatula or fork until a batter is formed. Be careful not to overmix! Fold in the zucchini.

Pour the batter into the prepared loaf pan and spread it evenly. Bake for 60 to 70 minutes, until a toothpick inserted into the center comes out clean. Remove from the oven and let cool for 20 minutes in the pan, then transfer to a wire rack.

While the loaf is cooling, heat a small, dry skillet over medium heat. Add the coconut flakes and sliced almonds and cook for 3 to 5 minutes, until a slight browning appears. Remove from the heat, let cool, and set aside.

When the bread has cooled, drizzle with agave nectar and sprinkle with the toasted coconut and almonds.

Piecrust in a Hurry

Homemade piecrusts can be intimidating, but with the help of our friend and colleague Jessica and her baking know-how, this recipe will save time in the kitchen. No need to prefreeze butter for this decadent crust; simply chill your vodka, add it to the mixture, and get your hands dirty.

YIELDS ONE SINGLE PIECRUST

1½ cups unbleached all-purpose flour

¼ teaspoon sea salt

½ cup Vegenaise

2 tablespoons ice-cold vodka

About ¼ cup ice-cold water

Preheat oven to 350°F.

Sift the flour and salt together into a medium bowl until well mixed. Add the Vegenaise and mix in with a pastry cutter or butter knives held between two fingers, until the dough is crumbly.

Add the ice-cold vodka and mix with your hands, then add 1 tablespoon of ice-cold water and mix again. Add another tablespoon of cold water and mix gently, being careful not to overwork. The dough should be just coming together, not too wet or tough. Use more water, adding 1 tablespoon at a time, if dough is still crumbly.

Use immediately. Roll out to your desired thickness and width between two pieces of plastic wrap. Press into a 9-inch pie plate, cut the excess from the rim, and shape the edges. Line the center with pie weights.

Bake on the middle rack of the oven for 25 to 35 minutes, until golden brown.

Summer Fresh Strawberry Pie

My family has always pulled over and stopped at roadside fruit stands to get the freshest organic strawberries and fruit varieties that you can't find in stores. What better way to use our Piecrust in a Hurry (page 196) than with amazingly fresh summer strawberries? Visit your local organic farmstand, get the best basket, and you won't regret it.

YIELDS ONE 9-INCH PIE; SERVES 4 TO 6

4 pints fresh strawberries, rinsed and patted dry

¾ cup evaporated cane sugar

2 tablespoons cornstarch

1½ teaspoons pectin

Pinch of sea salt

1 tablespoon fresh lemon juice

1 baked Piecrust in a Hurry (page 196)

Nondairy whipped cream for serving

Prepare the strawberries by hulling them with a corer or stainless-steel straw. Slice any large berries in half.

Select about 1½ cups of the imperfect-looking strawberries and puree in food processor or chopper until smooth, yielding about ¾ cup of puree.

Whisk together the sugar, cornstarch, pectin, and salt in a medium saucepan and stir in the berry puree. Cook over medium-high heat, stirring frequently with a heatproof spatula, making sure to scrape the sides and bottom of the pan. Bring to a boil, cook, stirring constantly, for 2 minutes. Transfer to a large bowl and stir in the lemon juice. Let cool to room temperature.

When the puree is cool, gently toss the remaining strawberries with the puree in a bowl to coat evenly. Carefully scoop the berries and puree into the baked piecrust. Chill for at least 2 hours before serving. Serve within 2 to 5 hours, topped with nondairy whipped cream, for the freshest flavor and texture.

Caramel Vanilla Cupcakes with Blueberry Pecan Topping

We doubled down and added fresh caramel filling to our classic vanilla cupcakes. Then, we muddled fresh blueberries with sugar and added it to our cream cheese to create a decadent frosting and it didn't stop there. We sprinkled them with crushed, toasted pecans because we felt like it and we'd like you to just sit back and enjoy these. They're a bit messy but, man, they're worth it.

YIELDS 12 CUPCAKES

Cupcakes

1¾ cups unbleached all-purpose flour

1½ teaspoons baking powder

⅛ teaspoon salt

½ cup Vegenaise

1¼ cups evaporated cane sugar

¾ cup almond milk

1½ teaspoons vanilla extract

Blueberry Frosting

½ cup fresh blueberries

⅛ cup evaporated cane sugar

½ teaspoon pure lemon juice

½ cup vegan butter

One 8-ounce container Follow Your Heart Dairy-Free Cream Cheese

2 cups powdered sugar

Caramel

½ cup vegan butter

¼ cup pure maple syrup

1 cup light brown sugar

¼ cup almond milk

1 teaspoon vanilla extract

½ teaspoon salt

½ cup toasted pecans, finely crushed

Prepare the cupcakes: Preheat the oven to 350°F. Line a muffin pan with 12 cupcake liners.

Whisk together the flour, baking powder, and salt in a medium bowl.

Whisk together the Vegenaise and cane sugar in a large bowl until creamy. Add the milk and vanilla and stir to incorporate.

Adding one-third of the flour mixture at a time, combine the flour mixture with the wet mixture and gently stir to incorporate, gently folding and taking care not to overmix.

Fill the cupcake liners two-thirds full. Bake 20 for 25 minutes, until a toothpick inserted into the center of a muffin comes out clean. Remove from the oven and let cool completely on a wire rack.

Prepare the Blueberry Frosting: Combine the blueberries, cane sugar, and lemon juice in a small saucepan over medium heat. Stirring frequently, simmer until the blueberries begin to soften. Mash with a fork or masher until most are blended and the liquid is reduced. Remove from the heat and set aside to cool completely.

Combine the vegan butter and cream cheese in a large bowl and cream together with an electric mixer on medium-high speed for 1 minute or whisk (with extra elbow grease) until well mixed. Add the powdered sugar, ½ cup at a time, and mix until smooth and creamy. Gently fold in the cooled blueberry mixture until evenly incorporated. Refrigerate for an hour if the frosting is too soft.

Prepare the Caramel: Melt the vegan butter in a small saucepan over medium heat. Add the maple syrup and sugar and whisk until dissolved. Bring to a boil, then lower the heat to medium-low. Add the almond milk and salt and whisk constantly for 10 minutes, being careful not to burn, lowering the heat if necessary. Add the vanilla and whisk to incorporate, then continue to cook, stirring, for only 1 minute. The color should be a rich golden, and the caramel should begin to stick to the sides of the pan.

When the cupcakes are completely cool, using a wooden spoon, poke a hole in the center of each cupcake and fill with caramel. Place the cupcakes in the refrigerator for 5 minutes to set, then frost and sprinkle with the pecans.

ACKNOWLEDGMENTS

For someone whose career is not centered around writing, the mere thought of creating a book can be daunting. So, when that book actually happens, one can only look around in amazement and gratitude at the people who brought it into being. My role, for the most part, was in the original creation of Vegenaise and subsequently the company that produces it. I never even considered writing a cookbook that portrays so elegantly some of the many ways it can be used. For that gift, I have to thank Danny Seo, who made it obvious that we need this cookbook and helped push us and connect us to the key publishing contacts that have brought us to where we are today. Once that spark was ignited, the amazing Katherine-Rose Franklin stepped up to embrace this project and nurture it through to completion. She compiled, edited, and tested recipes, and even added a few of her own. I honestly don't know who else could have done this.

There is always a group of essential people necessary to make a book happen. I want to thank Joy Tutela for taking our idea to a publisher that not only believed in our mission but shared our goals: The Countryman Press. Ann Treistman, Isabel McCarthy, Nick Caruso, Nicholas Teodoro, and the entire Countryman team, you have been a dream to work with.

Beauty is in the eye of the beholder, although in this instance, it is in the eye of the creators. I'd like to thank the core group of people who hung out with us for days to visually represent the recipes in this book. Alexandra Grablewski, you have an amazing eye for composition and light and the zen you brought to the shoot was truly aligned with our goals. Brett Regot, if through your presence alone we could reproduce what you do to make a dish really mouthwatering, we'd never let you leave us. Maeve Sheridan, when we first arrived and saw your extensive table of goods, we had no idea what it meant to be a prop stylist. To all of you and your teams, I am immensely grateful.

Additionally, I'd like to acknowledge and thank (in no particular order, and with apologies to those unintentionally omitted) the many chefs, friends, family members, and coworkers who have directly or indirectly contributed to this book: Oscar Mendoza, Stacy Michelson, Olivia Roszkowski, Jessica Hallstrom, Heather Bell, Jenny Engel, Tal Ronnen, Chad Sarno, Derek Sarno, Proof, Anne Gentry, Akasha Richmond, Dave Anderson, Gordon Smith, Ron Pickarski, Jason Stefanko, Greg Eisenberg, Adrienne DuBois, Herber Maldonado, Scot Jones, Tanya Petrovna, Jackie Poles Ran, Sara Farwell, Aaron Goldberg, Claire Goldberg, David Goldberg, Carly Alexander, Oscar Valencia, and a great many more who I don't have room to thank here that have generously shared their inspiration with us.

Finally, I want to acknowledge and thank my incredible wife Kathy, whose love and support means everything. She's the real secret to the sauce.

INDEX

Dedication

I dedicate this book to the three other original founders of Follow Your Heart: Michael Besançon, Spencer Windbiel, and Paul Lewin. We were four very different ingredients who, through some miracle, came together at the right moment in time and created one magical soup. Without each other, none of this would ever have happened. However, I would like to specially dedicate this book to Paul Lewin, my friend for a lifetime, by my side the whole way, whose support, inspiration, and love made it all worth doing.

For information about permission to reproduce selections from this book, write to Permissions, The Countryman Press, 500 Fifth Avenue, New York, NY 10110

For information about special discounts for bulk purchases, please contact W. W. Norton Special Sales at specialsales@wwnorton.com or 800-233-4830

Manufacturing by Versa Press
Book design by Nick Caruso Design
Production manager: Devon Zahn

The Countryman Press
www.countrymanpress.com

A division of W. W. Norton & Company, Inc.
500 Fifth Avenue, New York, NY 10110
www.wwnorton.com

978-1-68268-534-1

10 9 8 7 6 5 4 3 2 1